COSMIC ANGER

COSMIC ANGER

Abdus Salam – the first Muslim Nobel scientist

by

Gordon Fraser

OXFORD
UNIVERSITY PRESS

Great Clarendon Street, Oxford OX2 6DP

Oxford University Press is a department of the University of Oxford.
It furthers the University's objective of excellence in research, scholarship,
and education by publishing worldwide in

Oxford New York

Auckland Cape Town Dar es Salaam Hong Kong Karachi
Kuala Lumpur Madrid Melbourne Mexico City Nairobi
New Delhi Shanghai Taipei Toronto

With offices in

Argentina Austria Brazil Chile Czech Republic France Greece
Guatemala Hungary Italy Japan Poland Portugal Singapore
South Korea Switzerland Thailand Turkey Ukraine Vietnam

Oxford is a registered trade mark of Oxford University Press
in the UK and in certain other countries

Published in the United States
by Oxford University Press Inc., New York

First published 2008

British Library Cataloguing in Publication Data

Data available

Library of Congress Cataloging in Publication Data

Data available

Typeset by Newgen Imaging Systems (P) Ltd., Chennai, India
Printed in Great Britain
on acid-free paper by
Biddles Ltd., www.biddles.co.uk

ISBN 978–0–19–920846–3 (Hbk.)

1 3 5 7 9 10 8 6 4 2

Contents

List of illustrations

Introduction

> 'There is a tide in the affairs of men,
> which, taken at the flood, leads on to fortune.
> Omitted, all the voyage of their life
> is bound in shallows and in miseries'
>
> *William Shakespeare, Julius Caesar.*

In his painstaking biography of Primo Levi, Ian Thomson writes 'It is fantastically difficult to fashion a narrative out of the inchoate facts of someone's life'[1]. It is even more difficult in a biography of someone like Abdus Salam, whose life spanned separate themes and different worlds, making a traditional sequential chronicle impossible. Thus, and for another reason explained later, this book does not read like a diary.

Science underpins the whole of our world and governs our lifestyle, but – remote and difficult in its mathematical form – has become a neglected branch of culture. Scientists who change our view of the cosmos are overshadowed by celebrities who superficially appear to contribute more to the tides that affect the affairs of men. Abdus Salam was no such Cinderella scientist.

As modern science advances, it becomes increasingly inaccessible to outsiders, and to put Salam's contributions to physics in context needs a lot of groundwork. Although he will be remembered for the work in electroweak unification that led to his Nobel Prize, he also trampled across a wide range of research topics, not always successfully. He published 275 scientific papers, of which the *Selected papers of Abdus Salam* collection[2] includes 65. I cover only his most significant successes and failures, and refer to a small subset of the selected 65.

I have taken a much broader approach to his background as an Ahmadi Muslim in what was then British India. His assimilation into Western, and particularly British, life was so complete that his real lifespace was little understood by his European colleagues, as their widespread use of the meaningless first name 'Abdus' attests. To understand Salam needs an appreciation of his roots in the Islamic history of the Indian subcontinent, the tectonic movements that led to the creation of the Muslim nation of Pakistan, and his eventual excommunication as a heretic – not all Muslims would agree that Salam was the first Muslim to earn a Nobel Science Prize.

I first saw Salam in 1962 at Imperial College, London, when he introduced the inaugural lecture of his colleague and friend, Paul Matthews, recently promoted to Professor of Theoretical Physics alongside Salam. I was attending the lecture simply to kill time between the end of the undergraduate day and what would come later. London in 1962 was an interesting place for a 19-year-old: I was one of the few people to have heard the embryonic Rolling Stones. The door of the lecture hall opened and gowned figures swept in – Salam, Matthews, Kemmer. A student in front of me turned round and declared to nobody in particular 'They ooze brilliance!'.

My aimless undergraduate studies found a target. Several years later, I became a research student in Salam's group at Imperial College, but had little contact with him as he was then spending most of his time at his newly created centre in Trieste. When I moved to science writing, he was continually helpful and constructive, unlike many others, who seemed to think that constant carping would boost their reputation. I came to admire his writing, remarkable for someone whose mother tongue was Punjabi, and whose scientific talks were often largely incomprehensible. At scientific meetings and conferences, like many others he insisted on cramming too much material into the allotted time, so that it would qualify for inclusion in the published proceedings. Salam's prose was different: measured and stately. He drew on a profound knowledge of science history, world affairs, English literature, and Muslim heritage, together with his own sense of destiny, and an uncanny ability to choose the right synonym.

At a memorial meeting for Salam in November 1997 at the International Centre for Theoretical Physics, Trieste – Abdus Salam's creation – the centre was renamed in his honour. Salam was primarily a scientist, but science writer Nigel Calder recalled Salam's other motives: his controlled fury – 'cosmic anger' – about the injustice of a world where lack of opportunity can handicap even the most gifted students; and his indignation at the decline of science in the heritage of Islam. It was a 'wonderfully romantic story' that needed to be told, of a young lad from a market town in the Punjab who became a leader of science and earned a Nobel Prize, but who also emerged as a champion of the world's poor.

Acknowledgements and sources

Much important material is in the books *Ideals and realities, selected essays of Abdus Salam*, which ran to several editions and numerous translations. However, these anthologies include much repetition of episodes, viewpoints and anecdotes, even within each edition, and unfortunately their usefulness is often severely hampered by the lack of an index. Presentations at special events and at science-history seminars are another valuable source. Reading Salam's contributions to these, one senses that he was leaving messages for his biographers.

Salam was a very private man who, apart from his family, had few close contacts in whom he would confide. This remoteness was accentuated (for me) by material on his early life in Urdu or Punjabi. However, his family and colleagues have collected information in English on Salam's early days, much of it highly anecdotal, based on oral accounts. In these stories, Salam's school and undergraduate achievements are usually given more prominence than his Nobel prize. In tracing Salam's life, whether through the printed word or from those that knew him, certain episodes are repeatedly encountered, and have become embellished with retelling. This has spawned tributes and eulogistic websites that endow Salam with almost saintly status. On the other hand, the enemies of the Ahmadi sect to which he belonged produce some hostile invective. Some Salam anecdotes are corroborated in one direction, but refuted elsewhere. It is sometimes difficult to know who or what to believe. Part of the problem stems from Salam having been an avid storyteller who could spin a good yarn and refashion episodes to suit his needs: there are examples in this book.

For the history and sociology of the Indian subcontinent, I leaned on the authority of Percival Spear and on Peter Hardy's the *Muslims of British India*. Lawrence Ziring's *Pakistan in the twentieth century* was my primary guide to the turbulent politics of this new nation. On the scientific side, two key sources for the broad picture of twentieth-century physics are Abraham Pais' *Inward bound* and *The second creation* by Robert Crease and Charles Mann. Pais' book is a masterpiece, but is curiously telescoped the wrong way in time, his narrative accelerating and becoming more sketchy as time advances. Crease and Mann take a complementary tack, and flesh out their story with a wealth of contemporary detail.

As a result, the two books dovetail nicely together. (Crease and Mann also vividly recount several interviews with Salam.) Salam's important scientific work has been assembled in *Selected papers of Abdus Salam*, with a useful commentary by the editors – Ahmed Ali, Christopher Isham, Tom Kibble and Riazuddin. Jagjit Singh's 1992 biography[3] benefited from much direct assistance from Salam himself. However it was completed before Salam's death and, while full of interesting anecdotes, is sketchy on Salam's scientific and administrative achievements.

Another invaluable compilation, particularly for Salam's background and the early history of the Trieste Centre, is Alexis de Greiff Acevedo's 2001 Imperial College PhD dissertation 'The International Centre for Theoretical Physics 1960–1979: Ideology and Practice in a United Nations Institution for Scientific Cooperation and Third World Development'. People at Trieste who knew Salam – Luciano Bertocchi, Paolo Budinich, Anne Gatti, André-Marie Hamende, Mohammad Hassan, Seifallah Randjbar-Daemi – were unfailingly courteous and most helpful to me. Budinich, aged 90, one of Salam's few close associates, was an inspiration, and his book *L'arcipelago delle meraviglie*[4] provided another rich lode of information. Pioneer Trieste librarian Maria Fasanella built up a valuable collection of Salam material and memorabilia, freely available, and kindly supplied me with copies of key documents. George Thompson, Daniel Schaffer and Lucio Visintin also provided guidance and help. Anna Triolo Dodds unearthed Trieste archive photographs.

My thanks go to archivists Malcolm Underwood and Fiona Colbert at St. John's College, Cambridge; Anne Barrett and Catherine Harpham at Imperial College, London; and Anita Hollier for the Pauli collection at CERN, Geneva. Library staff at CERN were most helpful: David Dallman procured Alexis de Greiff Acevedo's thesis and made valuable suggestions; Marie-Jeanne Servettaz produced obscure books via interlibrary loan. I am indebted to Professors Pervez Hoodbhoy and Faheem Hussain for their recollections of Salam, for their advice, and for valuable assistance in Pakistan. Thanks also are due to: Nigel Calder for the initial suggestion for the book; Peter Tallack and Simon Capelin for helpful early hints; Hafeez Hoorani and Isobel Shaw for their knowledge of Pakistan and its affairs; B. A. Rafiq, former Imam of the London Mosque, and the UK Ahmadi Muslim Community office for advice and assistance with Ahmadi matters; Peter Harper and the University of Bath team for their precious archive catalogue; Alexis de Greiff Acevedo for sharing his Trieste knowledge and insight; and Hassan Shah for information and photos from Government College University, Lahore. Assistance with photographic material came from many directions: Ahmad Salam; Tom Johnson and Meilin Sancho at Imperial College, London; Jonathan Harrison and Clare Laight at St. John's College, Cambridge; Marcia Tucker at the Institute for Advanced Study, Princeton; the Information Department of the International Court of Justice, The Hague; and Audrey Charvieu of CAP Images, Divonne. Kezia Storr of PA Photos broke the Gordian copyright knot for Nobel Prize Award photographs. Frank Duarte provided biographical material on John Ward from *Optics Journal*. Simon Mitton provided a useful link with Fred Hoyle, Salam's influential undergraduate tutor, and read a chapter. A useful focus was an event at London's Imperial College on 7 July, 2007, organized by Michael Duff, Christopher Isham, Tom Kibble, Arttu Rajantie and Kellogg Stelle, marking the fiftieth anniversary of Salam's appointment as Professor of Theoretical Physics at the College, and at which I had the honour of being invited to speak.

Whenever I asked, Salam's former students and colleagues—scattered over four continents—were extremely helpful. Ahmed Ali, Ken Barnes, Robert Delbourgo, Michael Duff, Faheem Hussain, Fayyazuddin, Gordon Feldman, Christopher Isham, Pervez Hoodbhoy, J. D. Jackson, Tom Kibble, Peter Landshoff, Jogesh Pati, John Polkinghorne, Jacques Prentki, Qaisar Shafi, Ronald Shaw, John Strathdee, Raymond Streater, John Taylor, and Steven Weinberg all provided information, answered questions and/or read draft texts. Tom Kibble at Imperial College generously made available his own valuable collection of Salam material. Yuval Ne'eman in Israel pointed me in unexpected directions, but his eidetic memory sadly closed forever in April 2006, when the book was still at a very formative stage. Martin Veltman was another strong influence.

Above all, I must thank members of Abdus Salam's family – Louise Johnson, Abdul Majid, Aziza Rahman, Abdur Rashid, Ahmad Salam, Umar Salam – for their kindness, help, courtesy, patience and forbearance, especially when dealing with draft texts. I was fortunate to be able to interview Salam's wife Amtul Hafeez before she passed away in March 2007. Ahmad Salam's gracious hospitality at the family home in Putney and that of Louise Johnson at her riverside residence in Oxford are much appreciated. Sonke Adlung, my commissioning editor at

Oxford University Press, was a touchstone of wisdom and was able to summon impressive sources of authority, while his team performed the customary miracle of transforming an omnifarious package into a book.

Many biographers talk of it: now I know. Writing a biography is a lonely and bizarre business, sharing one's life with an intrusive figure who is nevertheless only a blurred shadow to the others who share one's life. My thanks to my wife Gill for enduring this predicament, and for being a continual sounding board for immature ideas; and to our children, Nathalie and Ben, for their valuable support, suggestions and continual interest.

Author's note

Archives are the raw material that a biographer should mine and refine. After Salam's death in 1996 and at the suggestion of his family, his papers and correspondence – 10 000 items in 350 boxes – were carefully collected and painstakingly catalogued by Lizzie Richmond, Paul Newman and Peter Harper of the British National Cataloguing Unit for the Archives of Contemporary Scientists at the University of Bath, UK. This work was organized from the International Centre for Theoretical Physics, Trieste, by Anne Gatti of the Director's Office, and funded by the Centre. The collected papers were then transferred to Trieste for safe keeping, with many folders subdivided into smaller files 'for ease of reference'. Unfortunately, Professor K.R. Sreenivasan, the Centre's Director since 2003, and who holds the title of Abdus Salam Research Professor, did not give permission to access these archives. No clear reason was given. This decision affected my approach, and anyone attempting to find here what Salam was doing on any particular day will be disappointed. In any case, Salam's achievements and failures are much more important than his often austere lifestyle.

Salam's papers at Trieste had been carefully divided into categories, one dealing specifically with the founding and administration of the International Centre for Theoretical Physics. However, the archive also contains simple press cuttings, and material covering Salam's personal history, and scientific research, much of which spans the years prior to the establishment of the Trieste centre. As Trieste had little experience of administering archives and making such material available to researchers, the Bath team offered some suggestions and guidelines.

Apart from a few items of evidently sensitive material, such as assessment of candidates for prizes, and elections to fellowships and positions, together with personal references and recommendations, the collection was seen as being freely available.

With the archives closed, the comprehensive six-volume, 840-page catalogue compiled by the Bath team thus became my precious guide to a forbidden territory. Much can be learned from such a detailed map. In previous years, researchers had not encountered this problem. Alexis de Greiff Acevedo's excellent thesis became especially valuable. My few explicit references to Trieste archive files are taken from this. Another fine source was André-Marie Hamende's consummate 'A Guide to the Early History of the Abdus Salam International Centre for Theoretical Physics, 1960–68', published from Trieste in 2002. However, I have not been able to access Salam's personal diary, which resides with the family.

A traditional pitfall is the transliteration of foreign names and titles. There are schemes that are meaningful to the initiated, but otherwise unhelpful (for example, Salam would be rendered as Salâm). I have simply followed common usage. The names of some places in the Indian subcontinent have changed over the years: in most cases I have retained the original name.

Gordon Fraser
Divonne-les-Bains
January 2008

REFERENCES

1. Thomson, I., *Primo Levi*, (London, Hutchinson, 2002)
2. Ali, A., Isham, C., Kibble, T., Riazzudin (ed.) *Selected papers of Abdus Salam* (Singapore, World Scientific, 1994)
3. Singh, J., *Abdus Salam, A biography* (New Delhi, Penguin, 1992)
4. Budinich, P., *L'arcipelago delle meraviglie*, (Rome, Di Renzo, 2002)

1

A turban in Stockholm

On 3 June 1925, Chaudry Muhammad Hussain had a dream. Normally, such dreams would quickly evaporate, leaving no trace, but this one had startled him, its sharpness etched in his memory. The date also lodged in his mind – it was an official holiday, the birthday of the Emperor of India, the British King George V. Otherwise, daily life in the market town of Jhang in the Indian province of Punjab continued much as it had done for centuries. Mains electricity and other modern prerequisites had not yet arrived. In the surrounding fields, gaunt oxen plodded in circles to drive wooden wheels that slopped water into irrigation channels. Unhitched from the wheels, the same animals pulled overloaded carts along potholed roads. In front of their mud houses, women cooked on open fires fuelled by dried cow dung, which they had moulded by hand into cakes and left to dry in the sun. There were a few creaking bicycles. In another rare sign that an industrial age had arrived, the British Raj had built a railway line through the town, connecting it with the city of Lahore, two hundred miles to the east, and the daily arrival of newspapers and mail at Jhang station was a major event. The waters of the mighty Punjab rivers had been channelled into vast irrigation schemes, but subsistence farmers still cultivated their smallholdings and looked after their livestock. In this humdrum world, Chaudry Muhammad Hussain was a local schoolteacher, instilling the rudiments of English and mathematics to boys in a classroom almost devoid of furniture. In the midst of such dreary predictability, the dream burned in his mind. In Islam, vivid dreams are not wisps of subconscious fancy, for it was in such visions that Allah revealed himself to Muhammad, who went out and changed the world. The Holy Qur'an, word of Allah as revealed by Muhammad, urges believers to search for understanding. 'It is he who drew forth the Earth and made it productive . . . . he causes the night to cover the day. In all this there are signs for people who reflect.'[1] Chaudry Muhammad Hussain's dream told him that his future child would reflect and interpret these signs as no other had done.

For Chaudry Muhammad Hussain, the family was the sole predictable means of social security. Even for a schoolteacher, existence was still hand-to-mouth. Benefits from the government or from the religious community were rare. Parents would bring up children, as many as they could, for infant mortality was high, and only strong children could care for parents in their old age. Older children would also look after the younger ones. Giving birth in mud huts, attended only by their mothers or sisters, many women died in childbirth or from subsequent complications during their traditional six-week confinement. Chaudry Muhammad Hussain, born in 1891, knew. His father, Gul Muhammad, had been respected in the community as a religious scholar and a *hakim*, or healer, but despite his medical knowledge had been unable to save his wife, who died when Muhammad Hussain (the honorific title of *Chaudry* was only acquired later, with status) was just a small child. Muhammad Hussain had been brought up with his brother Ghulam Hussain, 17 years older, in his uncle's house, where there were no other children. Muhammad Hussain's lonely life changed in 1922, when he married Saeeda Begum and had a daughter, Masooda, in 1924. When the girl was only six weeks old, Saeeda Begum died and Muhammad Hussain was alone again, but with a daughter to support. Although by now inured to sadness and deprivation, Muhammad Hussain, still only 33, was too young to remain a widower: his orphaned daughter did not thrive, and there was little help. After the Imam of his community activated a networking system, a message arrived from a distant family in Santokdas, 60 miles south of Jhang, members of the Pathan Kukezai tribe. Muhammad Hussain married Hajira Begum on 12 May 1925: the couple stayed with the bride's parents for the traditional 40-day honeymoon.

In his dream, Muhammad Hussain had seen that his child would be a boy, a first-born son who would bring glory to God and honour to his family. But he would not be a warrior or a rich merchant: his achievements would be through wisdom and intellect more powerful than any sword or amount of money. In a world rich in tradition but bereft of visual stimulation, Chaudry Muhammad Hussain recounted his dream to his family and his Imam, who were impressed by his vividness and emphatic detail. Muhammad Hussain immediately began to make preparations. In the dream, an angel had said that the son was called Abdus Salam, a simple but honourable name. When the child was born on 29 January 1926, Muhammad Hussain sought advice for a name and wrote to his religious leader, who recalled Muhammad

Hussain's account of the dream. 'When God has given a name, how can we interfere', he admonished.

The Islamic faith attributes Allah, the Almighty One, with ninety-nine names, whose recitation is a powerful litany, an oral rosary. In this act of devotion, each attribute of Allah is prefixed by the Arabic definite article, *al*, giving, for example, *Al-Aziz*, the Mighty One, *Al-Malik*, the King or Sovereign, *Al-Hakim*, the Wise One. In more ancient times, the appeal of these names – wisdom, honour, love, ... – had led many civilizations into creating a separate deity for each, worthy of targeted worship in its own right. But Islam's pantheon of 99 names underlines absolute monotheism, the awesome power and extent of One, and only One, all-seeing and all-embracing God.

The Arabic word '*abd*', meaning servant or slave, when combined with the 99 names of Allah, opens up 99 reverential possibilities for naming his subjects. There is no stigma attached to having one of these traditional names. On the contrary, they are most honourable. In Arabic, short vowels are implied rather than written, and in many cases are also dependent on the surrounding consonants, so that 'the Servant of the Mighty One' is *Abd-ul-Aziz*. The '*ul*' or '*al*' of the definite article belongs more to the second noun, the attribute of Allah. In the West, this article has been wrenched from its noun, and the name has been distorted to Abdul Aziz, implying an illogical first name, Abdul, and a second, Aziz. However, the real name remains the constructed noun pair, one of the 99 ways of saying 'the servant of Allah'. Another of Allah's attributes is *As-Salam*, 'the (Source of) Peace', so that one of the 99 honourable names is *Abd-us-Salam*, the 'Servant of Peace', the name Chaudry Muhammad Hussain's dream had provided for his first-born son. In this choice of name, not only the vowel of the definite article has been affected by its neighbouring consonants, but also the 'l' of the definite article has been modulated into an 's', giving a smooth elision. In the written form, the 'l' is preserved, showing the name as Abd-ul-Salam – *Abdul Salam*. Eventually, Chaudry Muhammad Hussain's other sons would also have simple honourable names – *Abd-us-Sami* (Servant of the All-Hearing); *Abd-ul-Hamid* (Servant of the Praiseworthy); *Abd-ul-Majid* (Servant of The Most Glorious); the twins *Abd-ul-Qadir* (Servant of the Powerful) and *Abd-ur-Rashid* (Servant of the Guide); and *Abd-ul-Wahab* (Servant of the Bestower).

Chaudry Muhammad Hussain also followed a Muslim tradition of not using inherited family names. Family names were invented because there are many more people than there are forenames. With short Arabic names, such as those grouped around the ninety-nine possible *abd* constructions, family ties can be indicated adding the word *bin*, or *ibn*, meaning 'son of', so that *Ibn Saud* signifies paternal links, redolent of the charming Russian custom of the patronymic (Andrei *Dimitriyevich* Sakharov), or deeper tribal affinities. When an Arabic-speaking man reaches maturity and has children, his name can also be embellished by adding *ab* or *abu*, meaning 'father of…', as in *Abu Bakr*, or *Ab-Rahim* ('Abraham'). Thus in 1980 the Moroccan Royal Academy embroidered on Abdus Salam's unadorned name, calling him *Abu Ahmad Abd-ul-Salam*, acknowledging Salam's eldest son, Ahmad. The following year, the Emirate of Kuwait chose instead to add the suffix *bin Hussain*, after Salam's father. In 1934, as part of his drastic overhaul of Turkish culture and traditions to amalgamate his new nation with twentieth-century Europe, Kemal Atatürk made surnames compulsory. Until that point, most Muslim Turks had only their given forenames – just a few had surnames or had acquired distinctive nicknames. In the resulting titular profusion, authorities tried to distinguish people further by adding places of birth or parents' names, but this only added to the confusion. The Turkish leader set an example: born simply Mustafa Kemal, he awarded himself the surname *Atatürk* – 'father of the Turks'.

Dual Arabic names such as *Abd-us-Salam* are single entities, inseparable, and splitting them into a 'first' name and a 'family' name is, strictly speaking, meaningless and a sacrilege. When Abdus Salam came to Britain, he discovered much that was unfamiliar and, to him, even bizarre. When naturally addressed as 'Mr Salam', he found it odd to be called by only half of his name. Later, when colleagues asked him whether he wanted to be addressed by his 'first' name, 'Abdus', his initial response was 'My name means "Servant of God" – would you like to be called "Servant" or "God"'. (The same colleagues, if meeting someone called Jones-Ford, would never think of calling him 'Mr Jones' or 'Mr Ford'.) Later, as Abdus Salam increasingly assumed a mantle of Britishness, he got used to people calling him by a totally artificial name, although it must have jarred. He had observed that Western people go to great trouble to give their children sophisticated names like Christopher, but in familiar speech promptly abandon the mellifluous name and use the curt monosyllabic Chris. Muslims living in close association with Westerners have learned to live with such habits, and following the example of Atatürk find it convenient to adopt western-style family names, inherited from one generation to the next. Abdus Salam's children use the family name Salam. However, to avoid repetition, the subject of this book is called by the single name 'Salam'. Although formally wrong, at least it is a meaningful word, unlike the spurious 'Abdus' frequently used by Western colleagues and friends. In his work, Abdus Salam was formally addressed as 'Professor Salam', preserving a basic duality. In later life, after the trauma of being formally excommunicated in his own country, he added the forename 'Muhammad' to emphasize his deep personal commitment to his religion.

His name is not the only problem. Abdus Salam was born in British India, and was a subject of George V, King of the United Kingdom of Great Britain, and Emperor of India. In 1947, what had been British India was torn into two new nations – India, with a majority Hindu population, but with no official state religion, and a new Muslim country, Pakistan. Salam became a citizen of Pakistan while he was an undergraduate at Cambridge University. It was a nation into which he was thrust rather than born. To refer to Pakistan as a geographical location before 1947 makes no sense, and to refer to a Pakistani as having been born in India requires clarification. But at any time in his life, Abdus Salam was always a child of a vast subcontinent, stretching over some 3000 kilometres, as far as from London to Istanbul – from north to south, and from west to east; proud of its own history, traditions and culture. Cut off from central Asia by the high mountain chains of the Karakoram range and the Himalayas to the north, and by desert to the west, the Indo-Gangetic plain has felt continual tides of invasion and migration through the sluice gates of the lower mountains to the north-west – the Hindu Kush, Kirthar and Suleiman ranges, with their valleys and passes.

Many biographies trace the lives of people with humble origins. Such modest backgrounds may be unfamiliar to some, or even most, readers, but they are nevertheless understandable. In the case of a Punjabi Ahmadi Muslim boy living in a two-roomed house with no electricity or running water, this is less obvious. (However, readers already familiar with the history and culture of the Indian subcontinent could enter the book at Chapter 4.) Going from one extreme of obscurity to another, Abdus Salam became an elementary particle physicist, a scientist whose business it is to work out the natural calculus of matter. So the book also has to explain our understanding of another very unfamiliar world, the microscopic quantum domain deep inside the atom.

His major scientific achievement, according to his Nobel Prize citation, was for 'contributions to the theory of the unified weak and electromagnetic interactions between elementary particles'. The force of electromagnetism can be harnessed to turn mighty machines. Less visibly but far more importantly, it is the interaction that holds together atoms, the smallest components of everyday matter. Abdus Salam showed how electromagnetism is linked to another subatomic force, the weak interaction, whose existence only became clear in the twentieth century. Much less tangible than electromagnetism, the weak interaction within the atom is no less important in the grand scheme of things. It provides the sparks that ignite the fuel of the Sun's nuclear furnace. It is ironic that when Salam was first taught about electromagnetism at school in Jhang, there was no electric light. All reading had to be done in the daytime or while crouched by an oil lamp. The teacher pointed out that to encounter the mystical phenomenon of electricity, his pupils should take the train to Lahore, several hundred miles away. Few did. And at that time, even in Lahore nobody knew of the subatomic weak interaction.

Charting the progress of science has many difficulties and pitfalls, and has to tread a narrow path between comprehensibility and scientific correctness. Tracing the development of a scientific idea is awkward. Scientists, even the best ones, are frequently confused and sometimes wrong, and progress can be erratic. Ambitious strides forward can overstep the mark. Milestone papers are usually written in a contemporary context that reflects current preconceptions and confusion, and can be difficult to understand in retrospect after all the confusion has evaporated. Even after this happens, vestiges of the original approach remain, memorials to a previous age. The conventional QWERTY keyboard reflects a design where adjacent mechanical typewriter keys were arranged so as to be less likely to get in each other's way and jam together. Keyboards are no longer mechanical, but the bizarre QWERTY key arrangement remains, a reminder of once-inadequate technology. So it is with theory. Much valuable scientific work also investigates unexplored openings only to find an intellectual impasse. There is no map into the unknown, and someone has to explore each possibility diligently. Many investigations report such fruitless searches, which remain in the scientific literature as 'No Entry' signs, warning subsequent investigators and pointing them to the right path.

The leap from a remote Punjab village to the forefront of twentieth-century science and technology is not the only theme in Salam's life. A deeply religious Muslim, his faith was a continual guiding light, his scientific research rewarding him with a clearer picture of Allah's design. For him, science was a form of devotion, his reverence to a higher power.

As a Muslim he was in a minority in Hindu-dominated British India, one that over centuries had carved out a fragile coexistence. But like a dormant volcano, this tolerance could be deceptive. One of the achievements of British rule in India had been to weave a diverse tapestry of races, religions and cultures, from the dark Dravidian-speaking peoples of the South to the Indo-Europeans of the North, and inhabiting such a wide range of climates, from the jungles of Kerala to the snowy ramparts of the Himalayas, into a dominion apparently more cohesive than several European experiments in nationhood. As the move towards Indian independence gained momentum in the mid-twentieth century, many held that this ethnic and religious diversity should remain a vital part of India. But even in the middle of the nineteenth century, some Indian Muslims had become sensitive to their status, always dependent on the goodwill of the majority and on their own acceptance of being a minority, and began to work instead towards a goal that eventually became a separate Muslim state. When Pakistan came into being in 1947, it ripped apart the flimsy patchwork fabric of the subcontinent. Voluntary migration became enforced exile, rekindling religious intolerance that had never been far from the surface. In the resulting bloodbath, half a million people were killed.

In their corner of the Punjab, Salam's family did not have to migrate in 1947. Their problems began later. They belonged to one of the smaller of the 72 sects of Islam, the Ahmadis, a minority within a minority. In Hindu-dominated British India, such an obscure sect had never been very visible. In the Punjab, conflict had been a triangular affair between Hindus, Sikhs and Muslims. But in the new Islamic Pakistan, with Hindus and Sikhs now having fled, the Ahmadis became more visible and were soon the victims of vicious intolerance. Branded as heretics by orthodox Muslims, many departed, Muslim outcasts from a Muslim country. One was Abdus Salam. As his achievements accumulated and his fame grew, he initially achieved recognition in his 'home' country.

But in the abrupt changes of government that came to typify Pakistan, the ruling regime became less tolerant of the minority sect, and in 1974 Salam was excommunicated along with all other Ahmadis. The only Pakistani to have earned a Nobel Prize, Salam is often unacknowledged in his own country, his name omitted and achievements not mentioned. There is even hostile censure. Many do not concur that Salam was the first Muslim to win a Nobel Science Prize. But the ultimate irony is perhaps that for those Ahmadis who chose to remain in India after the 1947 partition, life eventually would have been less traumatic. There, they are no worse off than any other Indian Muslim.

A proud international figure but a sad exile, Salam devoted most of his life to such injustices of the world. As a scientist, he saw that rich countries achieve growth through a complex infrastructure that, as well as simply educating people, promotes pure and applied research to catalyse new technology, and improve the quality of life. The underprivileged inhabitants of the Third World have problems just making ends meet. When they do not know where the next meal is coming from, long-term aims have to be shelved. Even with international aid, these countries can only make a meagre investment in development work to change their precarious predicament.

A thousand years ago, Islam had this infrastructure – it was the centre of world civilization, with famous universities in Cairo, Cordoba, Baghdad, ... But as this intellectual flame spread to renaissance Europe, it flickered and almost died in the Muslim world. Other, younger nations in today's Third World have no academic tradition at all, and their talent has to struggle. Gifted students have to emigrate to learn, and then frequently remain abroad to make the most of their talents, which are then lost to their home countries.

Salam was deeply aware of the glorious tradition of Islamic science, underlined by the message of the Holy Qur'an, and the fundamental role of science in knowledge (the Arabic word *'ilm* covers both). He was also deeply aware of the problems facing developing countries, but realized that any Third World nation could do little on its own to change this situation. Using his intellect as a weapon and the predicament of the Third World as a flag, Salam's lifetime mission was to battle against the cruel imbalance of global wealth and resources, artificial and man-made. In setting up his International Centre for Theoretical Physics in Trieste, Italy, under the banner of the United Nations, he had to overcome fierce resistance from developed countries who maintained that a centre for underdeveloped nations would itself be underdeveloped. But his creation became a magnet for students and researchers from all over the world, an oasis of learning and a role model for future ventures. He shared his 1979 Nobel Prize with the American scientists Sheldon Glashow and Steven Weinberg. If Abdus Salam had not been, then the 'theory of the unified weak and electromagnetic interactions' would still have happened. But there would have been less international focus for Third World scientific talent and there would have been more injustice in this world.

There are many worlds of difference between the dusty market town of Jhang, and Stockholm, the proud capital of Sweden. This contrast was very evident on 10 December 1979, when the recipients of that year's Nobel Prizes collected their awards from the King of Sweden. At the august ceremony, men usually wear formal dress and a white tie. One figure did not. Camera lights flashed as a stocky bearded man with a turban and gold *khusa* shoes, pointed like twin crescent moons, stood up to receive the prize for physics. Abdus Salam, the first Moslem to receive a Nobel science award, was wearing the same style of headdress that that his father always wore in the Punjab and that he had worn himself as a boy. It was a careful statement of pride and humility: pride that a son of such a modest Third World town should attain the highest accolade of science, and humility in remembering these modest beginnings. But over the years Salam had forgotten how to arrange the elaborate headdress, and initially had to be assisted by the cook at the Pakistani Embassy in Stockholm. However, the cook came from another part of the country, and knew only how to arrange the headdress in the local style. This did not suit Salam, who had to sit quietly and recall how he had tied it as a schoolboy (he had once worn the turban on a less formal occasion – at his children's school garden party, where he sat in a tent and pretended to be an oriental fortune teller, but here appearances had been less important). Salam was a man of many worlds – his own home and family, a market town in the Punjab, the quiet gardens and quadrangles of Cambridge, the fierce rivalry of academic research, ministerial-level diplomacy, and now the Nobel Prize. Normally he wore a heavy three-piece suit, wherever he was, whatever the weather, but wearing a turban and traditional dress in Stockholm epitomized the huge leap of his life's accomplishment.

At the Nobel ceremony, each prizewinner gives a speech, explaining his work and presenting it in a personal setting. Scientists have to explain their work frequently to their peers, and give the same lectures over and over again to generations of students. But at the Nobel ceremony they can afford to step back and present their work subjectively, explaining what it means to them as well as to others. Salam must have thought long about his life's journey before writing his presentation. To achieve his scientific ambitions, he had to move half-way across the world and live in a world that was initially unfamiliar to him, and probably still could be uncomfortable and surprising. Throughout history, students of ability have had to uproot themselves, leave their homes, at least temporarily, to live in great centres of learning. It is a traditional rite of passage. To learn, students must go to the teachers, and when they in turn become teachers, they must remain where the students come.

The first Muslim to receive a Nobel Prize in Stockholm (Egyptian President Anwar al-Sadat had shared the Peace Prize the previous year with Israeli Prime Minister Menachim Begin, but the Nobel Peace Prize is traditionally presented in Oslo), Abdus Salam was mindful that he was a scientific ambassador of Islam. Few people in Stockholm knew the impact that Islamic scholars had made on the progress of science. Salam, resplendent in his turban, saw himself as a Muslim gospeller, restoring nameplates in the pantheon of science that had become obscured or even illegible.

Islam, like the Universe itself, had begun from a vital spark in almost nothing. Muhammad, the prophet of Allah, died in Medina in 632 leaving no sons and no obvious successor, but had set in motion movements that would go on to change the face of the world. His initially small band of followers were galvanized by the energy and motivation of their inherited message. These followers who emerged from the desert were no longer an isolated band of raiders. Instead, they were tightly coherent, and this coherence spread to surrounding clans as the infant Islamic movement took root. In the vast spaces of the desert, the surrounding empires of Byzantium and the Persian Sassanids had been unprepared for movements across unguarded frontiers in the vastness of the desert. By surprise, insistence and astuteness, the message of Islam spread. Within a hundred years, a new Empire stretched from the Pyrenees in the west to Samarkand in the east.

This Muslim cloak was no bloodthirsty conquest. This would come later, as strife was imported from outside, and elements of the Empire became fat and complacent. But in the seventh and eighth centuries, the desert Arabs, the spearhead of the movement, were proud of their accomplishments and fiercely motivated by the message of the unity of God left by his prophet. Acutely aware of their insignificance in the face of Allah, and with their knowledge limited to oral tradition, as they moved into new territory they began to see a very different world. If they were confused, the Holy Qur'an guided them. The word of God told them that 'Only those of his servants endowed with the right knowledge and who can visualize the unity of the Creator by pondering of the diversity of his Creation, hold Allah in reverential awe.'[2].

With their culture embarrassingly empty, the followers of Muhammad adopted whatever they found as their own. Arabic began to supplant Greek as the international language of intellect, and new centres of learning grew up. In 641 the Muslim army arrived at Alexandria on the Nile delta, which under the Ptolemaic dynasties had become one of the world's great centres of learning. Consolidating themselves nearby in the new city of Cairo with its Al-Azhar mosque, the Muslims established the prototype of the modern university, a pole of learning that attracted able and motivated students from far and wide. Other universities were established in Baghdad, attracting students from central and southern Asia; and in Cordoba and Toledo in Spain (*al-Andalus*, the land of the vandals).

New intellectual figureheads inherited the message left by Aristotle, Pythagoras and Archimedes, and made their own, Islamic, contributions. But these had long been overshadowed by the subsequent European renaissance, and Salam wanted to salvage and restore their lost reputation and honour. Two of them – Abu Ja'far Muhammad ibn Musa al-Khwarizmi, born in Khwarizm, Uzbekistan, and Al-Kindi, from Kufa in Mesopotamia – were leading members of the scientific academy of Baghdad under the Caliph al-Ma'mun in the ninth century. Both men stoked the new furnace of Islamic intellectual development, integrating ideas from the west and from the east into Islamic learning. Building on the ideas of Diophantus in Alexandria in the third century, Al-Khwarizmi developed a new technique for solving mathematical problems –*`ilm-al-jebr* (algebra), the science of combination or association. Our word 'algorithm', a rule for calculation, is a corruption of al-Khwarizmi's name. Another legacy from this work was the concept of zero as a number, the absence of which had hindered the usefulness of mathematics. The word 'zero' comes from the Arabic *sifr*, sign, whence also our word cipher.

In his various presentations, Salam would proudly list other adopted Muslim concepts and technologies, often signposted by words with Arabic, Persian or Turkish roots: for textiles, 'taffeta' (woven) and 'damask' (of Damascus); in warfare, 'arsenal' (house of manufacture) and 'admiral' (commander); in chemistry, 'alembic' (still), 'alcohol'(powdered antimony) and 'alkali' (ashes); for food, 'saffron', 'sugar', 'syrup', 'sorbet'/'sherbet' (drink); cordovan and moroccan leather, and new technologies such as glassware, clockmaking, architecture and gunpowder[3].

When Napoleon Bonaparte invaded Egypt in 1798, he took with him a large group of scholars, '*La Commission des Sciences et des Arts*' whose job was to carefully document everything they found. Their monumental work, '*Grand Ouvrage sur l'Egypte*', ran to 7000 pages and sparked an awareness of ancient Egyptian culture that continues to this day. Bonaparte could have copied this idea from the Muslim warlord Mahmud from the Ghazni region of what is now Afghanistan, who in the eleventh century pushed eastwards across the Indus. On one of his incursions, Mahmud the Ghaznivid took with him Abu Rayhan al-Biruni, born in 973 in the same region of central Asia as al-Khwarizmi. Al-Biruni's masterpiece *Tahqiq-i-Hind* ('History of India') opened up a hidden world of Indian culture and geography. One of his revelations was a new system of numerals, much better suited to computation than the cumbersome Roman system. '*Whilst we use letters for calculation according to their numerical value'*, wrote al-Biruni, *'the Indians do not use letters at all for arithmetic'.* Al-Biruni's work led to the introduction of what we now know as Arabic numerals. A contemporary of al-Biruni was the great Persian polymath Omar Khayyam, now best remembered for Edward Fitzgerald's nineteenth-century translation of his quatrains, the 'Rubaiyat', but who made seminal contributions to algebra and geometry, and instigated a major reform of the Muslim calendar.

Other Muslim intellectual beacons illuminated the darkness. Alhazen (Abu Ali al-Hassan ibn al-Haytham), born in Basra in 965, is now known for his 'Treasury of Optics', eventually translated into Latin and published several centuries after his death. It rejected the Greek hypothesis that vision was due to rays emitted by the eye, and suggested instead that rays of light passing through a medium take the path that is most optimal, a suggestion that went on to be used in successive theories of gravity and all other natural forces. On the other side of the Islamic world, in Cordoba, a towering figure in the twelfth century was Abu Al-Walid Muhammad ibn Ahmad ibn Rushd, known better in the West as Averroës, whose milestone works on medicine and philosophy lived for centuries. His contemporary in Cordoba was the Jewish polymath Moses ben Maimon (Maimonides), who helped reconcile ideas from Greek, Jewish and Islamic thought, and eventually moved to Cairo to become physician to Saladin (Salah-ad-Din al-Ayyubi). When in 1236, the Christian monarch Ferdinand III of Castile retook Cordoba, it was a turning point in Spanish history, but the traditional openness that had marked Islamic rule continued for another two hundred years. Although some scholars preferred to move to Grenada or to Muslim lands further south, others stayed, and universities, still staffed with oriental scholars, continued to attract dedicated young students from the backward countries of Northern Europe, the Third World of the thirteenth century.

In his Nobel Prize speech, Salam illustrated the glory of Islamic science by highlighting one of these itinerant students, Michael the Scot (he had no surname!), who had left his native country to seek enlightenment. After learning at the new universities of Oxford, Paris and Bologna, Michael made his way to Toledo, where he learned Arabic and became immersed in the rich archives there. The legacy of the Ancient Greeks had been preserved in Arabic translations on the shelves of the Islamic universities, while more recent works covered important progress in medicine and astronomy. Michael the Scot set to work translating these Arabic books into Latin, a first step in a movement that would ultimately lead to the European 'renaissance', a rebirth of the Ancient Greek tradition that had been conscientiously preserved and furthered under Islamic rule, but would mark the dawn of the modern European era in all branches of learning.

Several times in his subsequent writings, Salam returned to the mysterious figure of Michael the Scot, and there can be little doubt that he identified his own career with that of the itinerant scholar who had combed Europe for knowledge 750 years before. Like Michael, Salam had come from a land with no recent tradition of learning, and, by his own efforts and good fortune, had transplanted himself into a new life. Most of these itinerant medieval scholars would have returned to their homelands after their studies, there to become teachers, doctors or priests, traditional career paths for the intellectually gifted. One of Michael the Scot's teachers, unimpressed with his pupil's progress, advised him to return to his homeland and take up sheep shearing and weaving the wool. But Michael the Scot was more ambitious. Although he never forgot his homeland, he was continually driven by fresh ambition and turned his sights elsewhere, travelling to Sicily, becoming physician and astrologer to Frederick II, the Holy Roman Emperor and most powerful monarch in Europe.

Frederick II, 'Stupor Mundi' – the World's Wonder – inherited a vast empire from his grandfather Frederick I ('Barbarossa'). 'Stupor Mundi' was a master politician and tactician, considerably enlarging the empire inherited from his family. His astute negotiation for the return of Jerusalem to European rule and being crowned its King in 1229 brought him immense prestige. Speaking nine languages and literate in most of them, Frederick was also a patron of science and learning, promoting the writing of poetry in early Italian and publishing a manual on the art of falconry. His numerous impressive achievements were overshadowed by excommunication by Rome, politically driven. To have such a distinguished patron was another huge stepping stone for Michael the Scot.

The history of the Punjab stretches back far longer than that of Western Europe: national history merges with ancient history. Punjab means 'Land of Five Waters', the Jhelum, Chenab, Ravi, Sutlej and Beas, tributaries of the mighty Indus, which appear like fingers on the hand of a vast plain, merging into the single arm of the Indus as it flows towards the Arabian Sea. In the sixth century BC, when Europe was struggling to emerge from the Bronze Age, Cyrus the Great, the Achaemenid Emperor of Persia, crossed the Hindu Kush, extending an empire that already reached as far as Asia Minor in the West. Herodotus (485–425 BC), the father of history, described how the new provinces, including Gandhara in the area that is now Pakistan, provided troops for the Persian armies in their battles against the Greeks. This marked the beginning of a wave of Persian influence in the Punjab that was to last for almost two thousand years, until in the nineteenth century English officially replaced Persian as the language of administration and commerce. An important centre in Gandhara was the city of Takshashila, better known by its Greek name of Taxila, a focus for nascent Buddhism and a cross-roads for the exchange of trade and culture between the Indo-Gangetic plain and all parts of the vast Persian Empire. In such traffic, many of the Indo-European words that have roots in Ancient Sanskrit could have found their way westwards.

Then came Alexander of Macedonia, who steamrollered irresistibly eastwards out of Europe, establishing six regional capitals named after him. After establishing the most remote of these, beyond Samarkand, in 327 BC he recrossed the Hindu Kush, this time turning south-east into what is still romantically called the North-West Frontier Province. Alexander's aim was to reach the furthest reaches of the Persian Empire. However, this boundary was not marked with a line on a map – it was more a blur that faded into the plain of the Punjab. Herodotus had supplied a tantalizing guide book, full of tales of ants that dug gold from the ground, and tigers with a sting at the end of their tail. Reaching Taxila, Alexander conferred with a group of philosophers, who have passed into history and literature as the Gymnosophists, the 'Naked Philosophers'.

Alexander soon moved on to new conquests and more battles, but Taxila remained garrisoned by Greek-speaking Asian troops, 'Bactrian Greeks'. Advancing 200 kilometres to cross his next river, the Jhelum (to him the Hydaspes) at Jalalpur, the ensuing battle was to be Alexander's last major victory. Pushing further across the Punjab, he crossed the Chenab and the Ravi, by now both swollen in the torrential monsoon rains. Bivouacked in the plains where now Lahore and Amritsar stand, Alexander's army was flooded, footsore, and plagued by snakes. With their weapons rusting, their supplies mildewing and their clothes and equipment rotting in the humidity, the soldiers grumbled that they had come far enough. All the territory of the former Persian Empire now lay behind them. With his army having come as far as its motivation and logistics allowed, Alexander finally doused his personal ambition and embarked on the Jhelum/Hydaspes/Indus for the coast. After a gruelling desert march across the Sind desert and Persia, he died at Babylon in 323, still only 32 years old.

Reading standard western literature, it is easy to get the impression that the arrival of Alexander the Great in the Punjab in 327 BC was the dawn of a new era, in the same way that Columbus 'discovered' the Americas that were already flourishing. But Alexander's passing was a mere episode in the turbulent history of the Indian subcontinent. Many more invaders were to come across the passes of the North-West Frontier Province. But for a long time Alexander remained the one who had come from farthest away, leaving a deep impression in the collective subconscious of the Punjab. The next Europeans to arrive in the subcontinent came by sea, and would change it far more than Alexander did.

In the nineteenth and twentieth centuries, gifted sons of the subcontinent made the journey in the other direction: Mohandas Gandhi, the spiritual father of Modern India; Jawaharlal Nehru, its first Prime Minister; and Muhammad Ali Jinnah, the *Qaid-i-Azam* or founding father of Pakistan. These were men who learned in Europe and then returned to mould the modern history of the subcontinent. Abdus Salam too migrated westwards, there to achieve most of his aspirations. But he could not return, and other ambitions had to be left undone.

REFERENCES

1. *Holy Qur'an*, Sura 13;3
2. *Holy Qur'an*, Sura 35:28
3. Vauthier, J. *Abdus Salam, un physicien* (Paris, Beauchesne, 1990), 58.
   For a fuller list of Arabic words imported into English, see http://www.al-bab.com/arab/language/lang.htm#words

2

The tapestry of a subcontinent

Salam's assimilation into Western, and particularly British, life was so complete that his background was little understood by his European colleagues. For most of those who called him 'Abdus', it remained unknown and irrelevant. As an Ahmadi Muslim in British India, Salam inherited a special demographic and cultural legacy, which, iceberg-like, invisibly but inexorably guided his destiny.

Its inner mechanisms hidden from view by its sheer size, Asia has been a vast piston, an engine of history. Violent expansions produced huge pressure waves that travelled vast distances before their energy was spent. It has been a chessboard for three of the world's greatest conquerors – Alexander the Great, Genghis Khan and Timurlane. Remote from any ocean and with few recognizable borders, the volatile politics and traditional transhumance in these regions created far-flung tides of natural migration. When these movements periodically ignited into war, savage hordes erupted to buffet the Near East and pound on the doors of Christian Europe. Less distant, the fertile Indo-Gangetic plain was a natural overspill for such movements, which surged through the defiles of the high Hindu Kush, and across the Khyber, Gomal and Khojak passes of the lower mountains to the South. The history of these migrations has become the cultural inheritance of the peoples of the Indus plain, who acknowledge them as readily as Britons might do William the Conqueror, or Americans the Pilgrim Fathers.

In AD 711, some eighty years after the death of the prophet Muhammad, and a thousand after Alexander the Great, a new invader appeared in the Indus valley, this time from an unexpected direction. The young Arab leader Muhammed Bin Qasim had marched south-east from Basra along the coast of the Persian Gulf. For Arab armies, deserts were less of an obstacle than they had been for Alexander, and Bin Qasim went on to cross the sands of Sindh and Baluchistan to reach the Indus delta, there striking northwards until his army reached Multan in the Punjab. But this was a false dawn of Islam in the Indus valley, as Bin Qasim was recalled to Baghdad.

While Islam continued to advance outwards from Arabia in all directions, it did not reach the Punjab again until 977, when this time Turkish armies from the Ghazni region of Afghanistan crossed the Indus. Turks from central Asia had frequently appeared on the north-west frontier, but now they carried the banner of Islam. The first major Muslim ruler here was Mahmud the Ghaznivid, who established his headquarters in Lahore. Behind him, Ghaznivid territory was swallowed up by Seljuk Turks, pushing Mahmud eastwards. His sacking of the temple at Somantha was a wound that was slow to heal and whose memory scarred the collective Hindu consciousness.

On an earlier incursion, Mahmud had taken with him Abu Rayhan al-Biruni, born in 973 in the same region of central Asia as al-Khwarizmi. Al-Biruni had worked as a court astronomer for several regimes until his talents were commandeered by the Ghaznivids. His detailed astronomical observations pinpoint his personal history. One of his major works – 'Shadows' – demonstrates how he carried out his astronomical observations, and covers many branches of mathematics – arithmetic and number theory, algebra, geometry, and planar and spherical trigonometry. A method for determining the hours for prayer was calibrated according to the Christian Byzantine calendar and angered Muslim orthdoxy. Al-Biruni stayed in India for about ten years, learning Sanskrit, translating its learning into Arabic and compiling his masterpiece *Tahqiq-i-Hind (History of India)*. Al-Biruni's work was an indispensable handbook for newcomers to a civilization that was already old.

Internal squabbles between local Rajput clans simmered until the end of the twelfth century, when Muhammad Ghuri, the next Muslim invader, arrived across the Gomal Pass and gained control of Sindh and the Punjab, including Lahore. The Rajput princes forgot about their rivalries and united against a new enemy, but after two battles the kingdom of Delhi fell in 1192. The western part of his empire did not long survive the death of Muhammad Ghuri, but what remained further east went on to become the powerful Delhi Sultanate, the capital of a succession of Muslim dynasties. It was during this era that Salam's Rajput forebears converted to Islam.

On the far side of the Hindu Kush, pressure was building up again as the Mongol Emperor Genghis Khan galvanized the clans of central Asia into an irresistible force. Mounted archers outran their opponents, while huge siege bows fired missiles deep into fortifications. Genghis also wielded terror as a weapon. Millions of Muslim men, women and children in central Asia were slain after their cities and villages surrendered pitifully. A hundred years later, Mongol cavalry again began to gather, this time converted to Islam. Emir Timur led his armies flamboyantly from the front despite limping from a war wound in the leg, hence the unflattering Persian title Timur Leng or Tamburlaine ('Timur the lame'). Scheming, ambitious and cruel, Timur's might and vengeance were frequently unleashed against neighbouring Muslim realms.

Returning after each campaign to his magnificent capital in Samarkand, Timur looked to new conquests, and in 1398 marched south across the Hindu Kush, following in the footsteps of Alexander the Great as he crossed the rivers of the Punjab, and finally swooping on an unsuspecting Sultan Mahmud in Delhi. The city became an inferno and a bloodbath, presaging another cataclysm that was to stain the history of the British Raj 450 years later. Laden with booty and driving columns of slaves, Timur headed back towards Samarkand. Behind him, Delhi lay empty and in ruins, and would have to await another emperor before regaining its glory.

Timur's genius was carried in his genes, but his grandson, Ulugh Beg (1393–1449), turned out to be a genius of a totally different sort. He transformed Samarkand from a military stronghold into a cultural centre, building the *madrasa*, the religious school that still dominates the grand Regestan Square. In 1428 Ulugh Beg began to construct a huge astronomical observatory. Before the era of telescopes, observatories had to be large to achieve precision measurements: Ulugh Beg's was over 50 metres in diameter and 35 metres high. Its precision, the equivalent of measuring the width of a pencil at a distance of more than a kilometre, demanded appropriate mathematics, and Ulugh Beg's trigonometrical tables went to eight decimal places and are still impressively accurate. His 'Catalogue of the stars', *Zij-i Sultani*, was among the first comprehensive updates of astronomy since Ptolemy's classic *Almagest* more than a thousand years earlier.

Five generations after Timur, the Asian migratory piston expanded again, and another wave of Turkish-speaking rulers pushed southwards. In the fourteenth century, the Mughals had been the first to feel the wrath and spite of Timur, but by 1504, these emnities had been resolved and the Mughal Emperor Babur, Timur's great-great-great-grandson, had spread his empire to Kabul. As cheerful and charismatic as Timur had been cruel, Babur began to look east at the ruin that had once been the proud Delhi sultanate. It was to be the beginning of the great Mughal Empire, a golden era in the history of India and of the world, bringing unprecedented wealth and culture to a region that until then had been a backwater of Asia. Despite the Mughals being Muslim invaders from the north, initially speaking Turkish, they created a culture that is still seen and admired as an apogee of achievement, and typically Indian. Although resurrecting the fear of Timur, Babur preferred to build vast gardens rather than piles of skulls, and his memoirs, the *Baburnama*, were later translated in English and inspired many generations of British colonialists.

While Babur's son, Humayun, was unable to consolidate what his father had achieved politically, he was a cultured man. He founded the school of Mughal painting, and loved astronomy, building seven audience halls, each named after a different planet, to span the week's public audiences. He fell to his death in 1556 after viewing Venus from his library roof. It was left to his son, Akbar ('The Great', 1556–1605), now speaking Persian rather than the Turkish of his forefathers, to assure the Mughal Empire of its place in history. Akbar's armies soon conquered Jaipur, Gujurat, Bengal, Kashmir, Orissa and Sindh, establishing a major realm that was now essentially Indian, rather than an annex to central Asia. This huge empire was ethnically and religiously mixed, but ruled by a small but recognizable minority. To do this, Akbar recruited promising local talent, often from Rajput ranks. These officers and administrators became junior partners, rather than servants, of Empire. The rewards from this service outweighed what could be expected from obstinacy and rebellion. It was a model that the British were to emulate later.

At the head of an empire characterized by its variety, Akbar made a bold attempt at unification that did not succeed. From his majestic new capital in Fatepur Sikri, consulting with Portuguese Jesuits, and with Hindus, Jains and Zoroastrians, Akbar announced a synthetic 'Divine Faith' (*Din Illahi*). It was not a success, and died with him, but other notions of a hybrid, syncretic religion had already taken root under Guru Nanak, the founder of Sikhism, a movement that would go on to play an especially important role in the Punjab.

After Akbar, under Jahangir (1605–27) and particularly Shah Jehan (1628–57), the Mughal Empire reached a peak of magnificence, with the construction of the Taj Mahal and other vast monuments and palaces. New forms of art and music, typically Indian, made their appearance.

Abdus Salam was proud of this heritage, pointing out that the Taj Mahal in Agra and St. Paul's Cathedral in London are good indicators of the level reached by their respective civilizations in the seventeenth century. Each was designed to reflect the status of their dominion and impress the toiling masses. St. Paul's is still grand and impressive, but few tourists come from across the world to have their photograph taken in front of it, or arrive at both dawn and dusk to admire and compare how the building reflects tangential sunlight.

Shah Jehan's son Aurangzeb (1657–1707), impatient at his father's self-indulgence, seized power, imprisoning his father in Agra. In later life, Aurangzeb became pious and ascetic, in stark contrast to the pomp of his father's reign, and was buried in the earth in a simple tomb. The splendour of the Mughal Empire had run its course, but the seeds of Islamic, rather than material, pride sown during Aurangzeb's reign were later to flower in a new awareness of faith. Until then, the Mughals were mighty Emperors who just happened to be Muslim. Aurangzeb instilled religious pride in a minority population who could not aspire to be aristocrats.

After periodic clashes between power rivals and the followers of the new Sikh movement, the tenth Sikh leader, Gobind Singh (1666–1708), fashioned the sect into an identifiable religion with its own fierce militancy. A poet as well as a general, Gobind instilled loyalty and courage. Although their religion incorporated Hindu and Muslim ideas, the Sikhs were neither, and remained proud of their independence. The sacred Sikh city of Amritsar in the Punjab, with its Golden Temple, provided the newest religion of the subcontinent with a focus and place of pilgrimage.

With the early history of the subcontinent dominated by pressures from within Asia, the 3000-kilometre oceanic coastline had largely insulated the south of the subcontinent. Arab traders had been attracted by spices, textiles and semi-precious stones. Later came Portuguese, Dutch, French and British merchant venturers, but initially there was enough coastline for them not to tread on each other's toes. However, the incessant warfare in Europe eventually spilled over into India, where British forces established supremacy in Bengal. Through military might and cunning ruses and gambits, the British convinced local Indian rulers that Britain was best suited to look after their external security and trade, and eventually ruled over much of the subcontinent.

However, the Punjab and Sindh, and the Pathan tribes to the north-west, remained fiercely independent. By agreement with Britain, the one-eyed Sikh ruler Ranjit Singh, the self-styled 'Maharaja of the Punjab', ruled over the territory to the west of the river Sutlej. Under this regime, many Muslims attained important local positions. After Ranjit Singh's death in 1839, the fragile peace with the British evaporated. The Sikhs crossed the Sutlej, but after a series of fierce battles, the British annexed the Punjab in 1849. Soon, other vast territories were annexed for the Crown. As they swept all before them, the smug British were increasingly heavy-handed.

British influence in India was not limited to mere territorial acquisition. In 1813, London had voted for 'the revival and improvement of literature and the encouragement of the learned natives of India and for the introduction and promotion of a knowledge of the sciences among the inhabitants of the British territories in India'[1]. It was the beginning of an impressive educational infrastructure. Initially this effort was geared towards classical Indian literature and Sanskrit, but in 1834 the writer and historian Thomas Macaulay, appointed to the Supreme Council of India, advocated more teaching in English and the promotion of European literature and science. His objective was to create a new class of Indian intellectuals who could think along British, as well as traditional Indian lines. As new schools and colleges were set up, English replaced Persian as an official language, and a new generation of Indian administrators appeared. This removed traditional opportunities for middlemen and brought in its wake uninvited Christian missionaries and mysterious innovations such as the telegraph. The increased British influence was seen by the masses as arrogant and authoritarian.

In this charged atmosphere, a minor incident in 1857 sparked a bloody conflict, traditionally described by the British as the 'Indian Mutiny' but seen by the Indians as the first move towards independence. Rebels stormed the British settlement in Delhi, where Bahadur Shah had been proclaimed Mughal emperor, the heir to the glory of Akbar. In a brutal backlash, British forces from the Punjab retook a ruined city. Bahadur Shah's banishment to Rangoon was a turning point in Indian history, particularly for the Muslims[2]. With the Emperor gone, the ties that had held India together began to fray, the traditional compensating mechanism between Muslim and Hindu was destroyed, and the way became open for extremism. With their figurehead saviour removed, the Muslim minority had to focus on their faith. New revivalist views of Islam, modelled on puritan Arabian Wahabism, took root, seeking out and excising acquired Hindu traditions. This made the British even more suspicious of Muslim intentions. 'After 1857, the heavy hand of the British fell more on the Muslims than the Hindus,' wrote Jawaharlal Nehru in his autobiography[3]. Aware of their unpopularity, more conciliatory Muslims tried to align themselves with British policy, but these rents in the fabric of India would eventually lead to the creation of the separate Muslim nation of Pakistan.

With the boundaries of their 'Raj' finally fixed, the British turned to material projects. Public works were a new foundation for British influence, with huge commercial and industrial schemes, and the development of irrigation, particularly in the Punjab. East of Jhang, between the Chenab and the Ravi rivers, the British built a vast canal system. Further afield, India's great cities became interlinked in one of the largest railway networks in the world. The hardware, shipped piecewise from Britain, changed the face of the country. The train replaced the bullock cart as the main means of travel, and castes found themselves thrust together in crowded carriages.

Administration of all this was overseen by the tiny Indian Civil Service (ICS), a strict meritocracy that recruited aggressively from the Victorian intellectual elite of Britain. After achieving a good university degree, a scholarship, a prize, or some other notable achievement, candidates would prepare for a stiff selection examination by studying Indian law, languages and history. For the hundred years before 1947, the 500 million or so population of India was administered by no more than about a thousand ICS officers at any one time. Absurdly high standards assured the quality of the young recruits and the initial success of this scheme, but the input concentrated on Oxbridge graduates who had studied classics. This was no handicap in the nineteenth century, but Abdus Salam later saw how this was incongruous in a twentieth century so dominated by technology.

Although in principle all administrative positions had been open to Indians after 1833, the ICS selection examination took place in English in England, with at one stage the upper age limit for candidates being 19, although subsequently increased. This effectively restricted ICS recruitment of Indian nationals to those who had been well educated in Britain. The examinations viewed the world through a British telescope. In 1863, the first Indian passed the ICS examination in England.

After an imaginative ruling giving Indian judges the right to try Britons was rescinded, disillusioned British ICS members helped found the Indian Congress Party in 1885, which would go on to become the political foundation of Indian independence. Dadabhai Naoroji, who in 1892 had become the first Indian member of the British parliament in London, convinced the government of the need to hold ICS examinations both in England and in India. Born in Bombay in 1825, Dadabhai Naoroji, excelled in English and mathematics (as Salam was to do), and became Professor of Mathematics at Bombay's Elphinstone College in 1852, the college's first Indian professor. After his political career in Britain, he became President of the Indian Congress Party.

Entry to the Indian Civil Service was a challenge that few in the subcontinent, and even fewer Indian Muslims, could aspire to, but had been clearly laid down for Abdus Salam by Muhammad Hussain. From the Punjab, it was the limit of Muhammad Hussain's vision, and it dominated the plans for his son's education. Enveloped by the curtain of this imposed goal, Salam was initially unable to see any further. Only after he had progressed a long way could he draw back an imposed screen and search for his own objectives.

The mixed races of the subcontinent led to a heterogeneous culture. Language in the Indus plain had been subject to many influences: the Turkik of the early Mughals had soon given way to Persian, but the influence of Islam led to many words being imported from Arabic. Urdu ('camp language'), with elements of Punjabi and Persian and written in Arabic script, had emerged under the Ghaznivids. It was subsequently taken to Delhi, where it attained a new status as a literary and court language in the sixteenth century. The Punjab, with its rich history and culture, retained its own colourful language, and Jhangochi, the form spoken around Salam's home town of Jhang, is reputed to be the oldest and purest.

The Shakespeare of Punjabi was the eighteenth-century poet Waris Shah, who left a rich legacy of literature, including a wealth of sayings and idioms, which have become an integral part of Salam's mother tongue. Waris Shah's most famous work is the romantic tragedy of Heer Ranjha, a tragic tale of two lovers that has strong parallels with Shakespeare's Romeo and Juliet – 'Two households, both alike in dignity; in fair Verona, where we lay our scene.' Instead of the Montagus and the Capulets, Waris Shah has the rival Ranjha and Siyal clans, and the action takes place in and around Jhang. It is as though Shakespeare had set Romeo and Juliet in Stratford-on-Avon rather than Verona. In the fairy tale, 'Romeo' is Dhido Ranjha, who falls in love with the lovely but unattainable Heer, a Siyal daughter, after stealing her reserved seat on the ferry-boat across the Chenab. Their love initially meets with fierce Siyal disapproval, but when Heer's parents eventually agree to a wedding, her callous uncle poisons her on the eve of the ceremony. Ranjha, summoned from Jhang, dies of a broken heart on her grave. The tale has been made into several films, and Heer's tomb in Jhang continues to attract tourists and travellers, in the same way as Juliet's balcony in Verona is always surrounded by sentimental admirers.

Urdu's bard was Mohammed Asadullah Beg Khan (1797–1869), better known as Ghalib, who chronicled the turbulent twilight of the Mughal dynasty and the bloody revolution/mutiny of 1857. His writing of 'oceans of blood' in Delhi affected the mood of people across the region. His style, next to that of Waris Shah, is probably the most important example of classical *ghazal* lyric. The young Salam, as an impressionable student of literature as well as mathematics, published articles on Ghalib's work.

The first modern writer from the subcontinent to achieve international renown was Rabindranath Tagore (1861–1941), from an intellectual Bengal family. Although chiefly a poet, he also wrote novels, plays and short stories, many of which were translated into English. In 1913 he became the first Asian to be awarded the Nobel Prize for Literature, an achievement that wrong-footed many European intellectuals. The first Nobel laureate born in India was Rudyard Kipling, who won the Literature Prize in 1907. His famous short stories and novels still provide a useful mirror of Indian life in the late nineteenth century.

While the literary tradition of the subcontinent is clearly signposted, its scientific development was less coherent. After its shining successes in the first half of the second millennium, the scientific tradition in Asia largely evaporated as the initiative passed to renaissance Europe. The Mughal emperors were more interested in monuments than science. However, Ulugh Beg in Samarkand and Humayun in Delhi had established an illustrious tradition in astronomy, whose flame flickered on. Maharaja Jai Singh in Jaipur built huge instruments at his Jantar Mantar observatory and, in a twilight of naked-eye astronomy in the mid-eighteenth century, added detailed corrections to astronomical tables.

At the beginning of the twentieth century, science began to re-emerge as the huge investment in higher education in British India that had begun half a century earlier began to bear fruit. Almost a thousand years after Al-Biruni opened the eyes of the west to the prowess of the subcontinent, the chrysalis into which Indian science had withdrawn showed new signs of life. However, any renaissance initially had been stigmatized by the tragic figure of Srinivasa Iyengar Ramanujan (1887–1920) – 'The Man Who Knew Infinity'[4]. Born in a poor Brahmin family in Madras province, this mathematical Cinderella was eventually 'discovered' by Godfrey Hardy at Cambridge. Soon, Ramanujan was working with Hardy in one of the great collaborations of modern mathematics, one whose heritage is still being exploited. However, by 1919 his health began to suffer and he returned to India, where he died of tuberculosis in 1920. The intellectual rags to riches story has become a classic saga[5]. Ramanujan's premature death was a stepping stone to fame, and his story became an inspiration for future generations of mathematicians from the subcontinent. One was Salam, whose first published contribution to science in 1943, at the age of 16, took over from where Ramanujan had left off.

The tragedy of Ramanujan was soon overshadowed by the triumph of India's first Nobel Prize for science, for Chandrasekhar Venkata Raman (1888–1970). After studying at the University of Madras, he went on to achieve a high ranking in the Civil Service examinations, and spent ten years as a government accountant in Calcutta. Continuing with scientific research in his spare time, in 1917 he became Professor of Physics at the University of Calcutta. Here, using simple spectroscopic equipment, he discovered in 1918 what became known as the 'Raman Effect', the scattering of Einstein's individual light 'bullets' (photons) by molecules. It was one of the first examples of the enigmatic quantum nature of the interaction of light, and earned Raman the 1930 Nobel Prize for Physics. Not only was he the first from the subcontinent to win a Nobel Prize for Science, but it was awarded for research that was actually carried out in India.

Another modern renaissance Indian scientist was Sir Jagadis Chandra Bose (1858–1937), a pioneer of radio-wave technology, but who was most famous for his work on the sensitivity of plants to heat and light. A student of Jagdish Chandra Bose at Calcutta was Satyendra Nath Bose (no relation, 1894–1974), who in 1924, working at the new University of Dacca, explained Planck's quantum theory of radiation entirely from first principles. In 1900, Planck had put forward his revolutionary quantum theory of radiation as an empirical idea. It worked, but nobody knew why. Bose boldly mailed his explanation to Albert Einstein. Despite being bombarded by requests of all kinds, Einstein took the time to study the strange unsolicited paper from far away. It showed how Planck's law followed if bullet-like photons could collect together in the same quantum state. Einstein translated the paper into German and ensured its publication, introducing a new quantum principle which became known as 'Bose–Einstein statistics'. Particles, like photons, which can collect together in the same quantum energy state, are now collectively known as 'bosons', and they were to be an integral ingredient of Abdus Salam's new physics.

A nephew of Raman was Subrahmanyan Chandrasekhar, born in Lahore in 1910, but whose family moved to Madras in 1920. Influenced by his uncle and inspired by the legend of Ramanujan, Chandrasekhar became absorbed in the physics of stars, using mathematics to track what happens in fiery conditions and on astronomical scales that are far removed from those of any terrestrial laboratory. In 1930, he left India for postgraduate research at Cambridge, and on the boat had an epiphany that would change his life and the course of astrophysical thinking.

The inner thermonuclear furnace of a star, stoked by nuclei fusing together, gradually changes its composition and will eventually cut out when its supply of fusible material is exhausted. With the star's furnace extinguished, its internal pressure drops, and the star will shrink under the crush of its own gravity. Chandrasekhar realized that if such a star is large enough, it reaches what is now known as the 'Chandrasekhar limit', and its gravitation eventually implodes even its constituent atoms, forcing the atomic electrons into the nucleus, and the star collapses completely. In 1983, he shared the 1983 Nobel Physics prize for 'his theoretical studies of the physical processes of importance to the structure and evolution of stars'. Chandrasekhar worked at the Yerkes Observatory of the University of Chicago, about 150 kilometres north of the city. He taught astrophysics to many generations of students, calling for a long drive to and from the University campus. In the late 1940s, he lectured to a class composed of just two Chinese postgraduate students – T. D. Lee and C. N. Yang – who went on to share the 1956 Nobel Physics prize. That two young scientists from South Asia could achieve so much so rapidly was to be a major motivation for a third.

The most influential twentieth-century Indian scientist was Homi Jehangir Bhabha. Born in Bombay in 1909, he learned physics at Cambridge where, in the 1930s, he was the first to write down the quantum mathematics of a process now known as 'Bhabha scattering'. When he returning to India just before the Second World War, the powerful Tata dynasty helped him found the Tata Institute of Fundamental Research, formally inaugurated in December 1945. With talent pouring out of the universities, this soon became an additional nursery for India's substantial domestic science effort. Bhabha was President of the United Nations Conference on the Peaceful Uses of Atomic Energy, held in Geneva in 1955, where Salam was busy as a scientific secretary. In India, Bhabha became the figurehead of the national Atomic Energy Commission, which soon embraced the challenge of building an atomic bomb. Bhabha died in an air crash in 1966, well before India's first nuclear device was detonated in May 1974.

As the Indian subcontinent rushed to cast off the mantle of British rule, the rich tapestry of race, language and culture that the British had helped weave into an emergent nation was ripped apart into two independent states: India, with a majority Hindu population; and Pakistan, a home for the subcontinent's Muslims. The new India would continue the modern renaissance scientific tradition of Ramanujan, Raman, Bose, Bhabha and Chandrasekhar, but in Pakistan only a solitary initial figure was to emerge.

REFERENCES

1. Spear, P., *History of India*, Volume 2, (London, Penguin, 1990)
2. Dalrymple, W., *The last Mughal* (London, Bloomsbury, 2006)
3. Nehru, J., *Toward freedom* (Oxford, OUP, 1936)
4. Kanigel, R., *The man who knew infinity, a life of the genius Ramanujan*, (New York, Scribners. 1991)
5. Kanigel, 1991, see also Hardy, G. H., *A mathematician's apology*, (Cambridge CUP, 1992)

3

Messiahs, Mahdis and Ahmadis

Science was but one cornerstone of Salam's life: his faith was another. Religion, although firmly anchored in custom and dogma, is nevertheless always in transition. Even the oldest traditions and the most unshakeable dogma can be buffeted by world events, and the foundations of belief slowly, almost imperceptibly, evolve. New outlooks and interpretations continually reconcile religious demands with the evolving view of the world. The sixteenth-century Reformation and the emergence of Christian Protestantism in Europe is an example, while the eighteenth-century orthodox Wahabi movement in Arabia shows that reform can move both towards as well as away from orthodoxy. Some of these seismic shifts can be catalysed by a single event or person; some are more a product of gradual evolution; others are a mixture of the two. Three great monotheistic religions – Judaism, Christianity and Islam – although now very distinct, share a common heritage, in particular the concept of a Promised Messiah. One of the most recent examples of a perceived messianic fulfilment is that of the Ahmadi sect of Islam to which Salam belonged. While firm belief in a nineteenth-century Promised Messiah is merely curious to some, it is outright heresy to others. As well as being a driving influence on his personal and family life, Salam's fervent belief affected his status in Pakistan: his reputation in his home country was always a barometer of Ahmadi status.

The three monotheistic religions share a treasure of biblical literature. In this, the book of Isaiah stands out by its magnificent allegory as well as its length. In about the eighth century BC, Isaiah was part of the aristocracy of the Kingdom of Judah at a time when Assyrian armies threatened to invade. His evocative and sparkling prose – 'they shall beat their swords into ploughshares and their spears into pruning hooks' – has been borrowed by many languages. At a time when the spoken word ruled, such vivid and picturesque metaphors had a distinct advantage.

As well as denouncing the continual wars and battles that ravaged the region, Isaiah advocated social justice, despite his own aristocratic background, and had a broad outlook on contemporary politics. From this higher viewpoint, he also attempted to peer into the future, and proclaimed the eventual arrival of a saviour, a Messiah. The word, now freely used in many languages, comes from the Hebrew 'he who is anointed', pointing back to the biblical King David. The word was eventually translated into Greek as 'Christos', but in doing so took on a new, deeper implication.

Isaiah foretold: Therefore the Lord himself shall give you a sign; Behold. A virgin shall conceive, and bear a son, and shall call his name Immanuel *[from the Hebrew 'God with us']*. (Is 7)

He continued: And there shall come forth a rod out of the stem of Jesse *[the father of David]*, and a branch shall grow from his roots; And the spirit of the Lord shall rest upon him, the spirit of wisdom and understanding, the spirit of counsel and might, the spirit of knowledge and of the fear of the Lord;

And shall make him of quick understanding in the fear of the Lord: and he shall now judge after the sight of his eyes, neither reprove after the hearing of his ears;

But with righteousness shall he judge the poor, and reprove with equity the meek of the earth: and he shall smite the earth with the rod of his mouth, and with the breath of his lips shall he slay the wicked.

The wolf also shall dwell with the lamb, and the leopard lie down with the kid; and a little child shall lead them.

The earth shall be full of the knowledge of the Lord, as the waters cover the sea. And in that day shall there be a root of Jesse, which shall stand for an ensign of the people; to it shall the Gentiles seek, and his rest shall be glorious. (Is 11)

With these powerful words, the scene was set for the coming of a Messiah, thereby anticipating the problem of how to recognize him when he came. Was Isaiah's prophesy literal or metaphorical? History is full of attempts to reconcile the episodic appearances of purported Messiahs with Isaiah's exacting scenario.

Seven centuries after Isaiah, Jesus (Joshua, Isua) of Nazareth galvanized his small but faithful band of followers in Galilee, a backwater of the Jewish world. By this time, Isaiah's picture of an all-knowing Messiah had developed into one of a political and military firebrand who would save the downtrodden Jewish people. Jesus did not conform to this picture. The contemporary Jewish orthodoxy in Jerusalem did not see in him the militant Messiah they had been led to expect, or even a prophet. Christianity, initially a splinter sect of Judaism that acknowledged a known Messiah, held that Jesus, the incarnation of the Son of God, died to atone for the sins of humanity. After an evocative martyrdom, his spectacular resurrection convinced his followers to spread the gospel. The emblematic cross on which he died became a rallying cry for a new religion, and a ready target for its opponents.

Christianity grew in popularity in a world strongly influenced by Greco-Roman culture. It received a major boost after the revolt of the Jews against occupying rule in 66 AD, the subsequent brutal Roman invasion, and the savage destruction of Jerusalem. In the ensuing diaspora, a new message was distilled from the context of Judaism and around the figurehead of Jesus Christ. Stripped of constricting observances such as circumcision and strict dietary laws, the dynamic new faith blossomed. All who wanted and were willing to accept its tenets were eligible.

The Jews, uncomfortable after the Christians had their Messiah, resigned themselves to wait for their own. This became more difficult after the destruction of the Temple in 70 AD and the subsequent Jewish migrations. Sporadic claims were made in various places, usually dealt with promptly and brutally by the local authorities. In the seventeenth century, after the Jews had been expelled from Spain, the charismatic Shabbetai Tsvi was initially widely heralded as the Messiah, but, like Jesus Christ 1600 years previously, was viewed sceptically by Jewish authorities in Jerusalem, and even more so by the Ottoman Sultan in Constantinople. He faded into obscurity, but his legend bred a series of occult Shabbetai Tsvi lookalikes, which confused onlookers who had previously been tolerant of Judaism, bringing a new era of persecution. In a very different setting, the twentieth-century Emperor Haile Selassie of Ethiopia became a messianic figure for the Jamaican Rastafari movement.

Six centuries after Christ, the Holy Qur'an, the word of Allah, was freshly relayed by his prophet Muhammad. A new religion, Islam, burst out of Arabia onto an unsuspecting world. Its compelling message soon blanketed a vast area, bringing its own vision of the Messiah. While the name has periodically evolved to imply a person of great power, the word in its original context of a descendant of David is much less evocative. The Holy Qur'an says of Jesus:

> Allah gives you good tidings about a word from Him on the birth of a son whose name is the Messiah, Jesus, son of Mary, he shall be worthy of regard in this world and the hereafter. [Sura 3.45]

Jesus was messianic, but the Holy Qur'an, the word of Allah as revealed to Muhammad, reacted to the now widely established Christian credo and redefined the role:

'The Messiah, son of Mary, was only a messenger, like all the messengers who have passed away before him'. [Sura 5.75].

Islam refuted the divinity of Jesus. Orthodox Muslims believe that Jesus, spared by God from death on the cross, ascended bodily to heaven, and will eventually return to relay his message and purify the faith of the believers:

They [the people of the Scripture] claim "We did kill the Messiah, Jesus, the son of Mary, the Messenger of Allah", whereas they killed him not, nor did they cause his death by crucifixion, but he was made to resemble one who had so died [Sura 4.157].

To provide a robust saviour, Islam introduced the *Mahdi*, the 'rightly guided one', or redeemer, who would usher in a glorious new age and 'fight' alongside the returned prophet Jesus. Shi'a Muslims believe in a 'Hidden Imam' and Mahdi. In 874, after the death of the incumbent Imam, the seven-year-old Muhammad al-Mahdi declared himself to be the next leader, but went into hiding. Shi'a Muslims pray for the return of the Mahdi who will set humanity on a new path.

The importance of a Mahdi increased as the once-mighty Ottoman Empire, the political heir of Islam, began to crumble in the nineteenth century. A whole litany of attributes developed to be able to recognize him. One of the most famous contenders was Mohammad Ahmad bin Abdullah (1843–85) who used a powerful Islamic message to rally a revolt against Egyptian rule in the Sudan. Here was a Mahdi who wanted to fight with the sword. His army besieged Khartoum in 1884 and killed the British governor, Charles Gordon, who had been appointed by the Egyptian Khedive. Mohammad Ahmad died soon thereafter, and the residual rebellion was crushed at the battle of Omdurman. His brief period of glory echoed round the Islamic world.

This episode was a reaction to European colonial rule. A very different contemporary example came elsewhere in the British Empire, in Qadian in the Gurdaspur district of the Punjab, some 40 miles east of Amritsar, the spiritual centre and focus of pilgrimage for the Sikh religion. In Qadian, Mirza Ghulam Ahmad (1835–1908) was to proclaim himself Messiah and Mahdi. Like other proclaimed messiahs, his early life was unspectacular. Dominated by his influential father, he was an overseer of the swathe of agricultural land allotted to the family by Ranjit Singh. He went into government administration at Sialkot, but spent most of his time studying the Holy Qur'an. After his father died in 1876, Mirza Ghulam Ahmad's life suddenly changed course.

The growing awareness of Indian nationalism and the need to reconcile it to a plurality of local religions had seeded new movements. Ramakrishna had emerged as the figurehead of a more modern Hinduism. This was countered by a revivalist Hindu movement, *Arya Samaj* ('Society of Nobles'), which called for a return to the fundamental principles of the basic Vedic hymns, abandoning much accumulated tradition. A powerful missionary movement transmitted the *Arya Samaj* message, emulating the efforts of Christian missionaries in India. This rode on the contemporary wave of emigration of Hindu workers to British territories in Africa and the Pacific. In the Punjab, *Arya Samaj* also focused in the Sikhs, emphasizing their roots in Hinduism. Convinced that Islam too should be dynamic, Mirza Ghulam Ahmad aimed to restore the faith of Muslims who had lost confidence in the face of the energetic efforts of Christian missionaries and *Arya Samaj* Hindus.

Only a hundred miles from the Punjab, in Srinagar in Kashmir, is the tomb of an ancient prophet, Yus (Jesus) Asaf, linked by local legend to Jesus of Nazareth, who had said 'I am not sent but unto the lost sheep of the house of Israel' (Matthew 15.24). A terrestrial resting place for Jesus contradicts the tenets of both Christianity and orthodox Islam. According to the legend, after recovering from his torture, Jesus fled the troubled area of his birth and eventually settled in North India. (Other contemporaries of Jesus travelled equally far: St. Thomas – 'Doubting Thomas' – journeyed to the Malabar Coast of Southern India, where he founded a strong Christian tradition that used the Syriac liturgy; when Jews fleeing the Roman destruction of the Temple arrived in the area of Cochin some forty years later, they reported a Christian community already established there.) In North India, Yus Asaf lived out his life and died at the age of 120. His descendants thrived, and by the end of the nineteenth century had reached the sixty–fifth generation.

In 1876, Mirza Ghulam Ahmad was 41, a propitious age, the same as that of Muhammad when the Holy Qur'an had been revealed to him. It was also a propitious time in the Punjab. Mirza Ghulam Ahmad went out and proclaimed himself the heralded Messiah and Mahdi. Based on divine revelations, he declared that he resembled Jesus in face and stature, and had been sent to 'break the cross', and show crucifixion to have been a fable. The revelations displayed that the teachings of the Holy Qur'an were still as relevant as they had been 1400 years previously. Muslims, leaderless after the Mughal emperor had been banished to Rangoon in 1857, were searching for a new standard around which they could rally. Some had looked west towards the traditional home of Islam, and a groundswell of revivalist orthodox Arabian Wahabism became popular, worrying the British. The warlike tradition of *jihad*, holy war, was never far from Muslim consciousness. But Mirza Ghulam Ahmad pointed out the folly of any jihad against a British rule that had protected Punjab Muslims from the sword of Ranjit Singh's Sikh armies. He was a Mahdi whose jihad would rely on reasoned argument, not military might. This pacifist credo, combined with the rekindling of interest in fundamental Islamic values, boosted the popularity of the movement. Endowed with the title of *Hazrat*, Mirza Ghulam Ahmad went to the people, and they listened. His energy convinced his followers that they were witnessing an Islamic renaissance, and his message assuaged confusion. His followers became seized by missionary zeal, and copied modern Christian and *Arya Samaj* techniques to spread the Ahmadi word, which quickly became popular in West Africa.

Christianity had never been an official part of British policy in India. On the contrary, the new educational infrastructure of the mid-nineteenth century set out to complement, rather than to replace, the culture and religious beliefs of the subcontinent, producing educated people who could think along European and Asiatic lines. But Christian missionary zeal was never far from the surface. In Africa, the campaign against the slave trade had initially focused missionary attention on that continent. In India, incomprehensible customs could also become the incarnation of evil – notably the Hindu tradition of a widow's self-immolation on her husband's funeral pyre. Lurid tales of murder, rape and other atrocities during the revolution ('Mutiny') of 1857 launched a paroxysm of British revenge in the short term, but also continued to support the missionary zeal that had helped generate the revolution/mutiny in the first place. Although not part of official policy, 'Onward Christian Soldiers' was nevertheless an apt anthem for India in the second half of the nineteenth century. The well-supported London Missionary Society amassed large sums of money to equip and maintain missionaries there. Under the mantle of British rule, Christian missionaries were active all over the country, especially as teachers, working in Christian colleges where Indian students would assimilate Christian ideas during their education. Itinerant missionaries visited remote villages, and Abdus Salam's maternal grandfather, Nabi Bakash, in Batala, north-east of Amritsar, troubled by their persistence and their message, sought the advice of Mirza Ghulam Ahmad (who had gone to school nearby).

The prominent role of Jesus/Yus in the Ahmadi credo could help puzzled villagers to reconcile the unfamiliar message peddled by Christian missionaries with their knowledge of Islam. Christian missionaries unwittingly provided fertile ground for the Ahmadi message. But not all were convinced. Mirza Ghulam Ahmad's startling claims and changed perception of the 'crucial' role of Jesus easily offended Christians and orthodox Muslims alike. His pretensions also puzzled many who had known him all their lives. If he was on Earth for a specific purpose, why had this been hidden for so long? His insistence inevitably irritated and then enraged Muslim orthodoxy. Apart from the refocused picture of Jesus, their main objection was to his claim to be a visionary, and in some sense a prophet. This contradicted a central tenet of Islam – the absolute finality of Muhammad's prophethood. Hazrat Mirza Ghulam Ahmad patiently explained that he was not a law-bearing prophet, simply a 'reflection' of the Holy Prophet who refocused believers' attention, rooting out accumulated corruption and pollution. He was an interpreter of the law, not its bearer. But such a claim was nevertheless sacrilegious to mainstream Islam, which saw him as a false prophet, as the Jews had viewed Jesus nineteen hundred years before.

Initially, the Ahmadi movement had been indiscernible, but as numbers swelled, an orthodox backlash tried to nip the bud of the new sect. Their pro-British stance also angered Indian nationalists. Ahmadis were discriminated against, persecuted, barred from praying in mosques, and their funerals prevented from using Islamic burial grounds. Bodies were disinterred, and traditional Muslim inscriptions on Ahmadi tombstones routinely defaced with hammer and chisel by those with a grudge. There were, and still are, allegations that Mirza Ghulam Ahmad's followers, as part of their openly pro-British stance, passed on information about dissidents and seditioners[1]. After the establishment of the British mandate in Palestine in 1923, it was easy for the opponents of the Ahmadis to link anglophile Ahmadi policy with the growing awareness of Zionism, given new impetus by the 1917 Balfour Declaration favouring the establishment of a Jewish home in

Palestine. 'Hostility towards Islam and hatred towards Muslim Ummah forms the bed-rock of intimate attachment between Qadianis and Jews.'[2] The existence of an Ahmadi mosque in Haifa, Israel, provides another convenient target.

In modern Pakistan, Ahmadis are called Qadianis, after Qadian, the home town of Mirza Ghulam Ahmad. But after the partition of the Punjab in 1947, Qadian found itself in India, so the headquarters of the movement moved to Rabwah in Pakistani Punjab, which became a thriving Ahmadi community. (In a further official slap on the face, the authorities renamed Rabwah as Chenab Nagar.) A tiny population remained in Qadian to look after the tomb of their Messiah and Mahdi. Education is considered especially important by Ahmadis, a stepping stone to advancement: the literacy level among Pakistani Ahmadis is almost 100%, compared with a national average of around 50% (higher among men than women). There could be about ten million Ahmadis throughout the world, with about four million in Pakistan. Castigation is a fate shared by many upstart religious movements, their adherents shunned and ostracised. Ahmadis are no exception. Like the Jews or the Pilgrim Fathers, many of them uprooted their standard and planted it in new homelands, there to thrive by their own industry and effort.

Abdus Salam's father, Muhammad Hussain, while studying at Islamiyya College in Lahore in 1914, had been asked to help produce some anti-Ahmadi literature. This troubled him, as he had no personal quarrel with the movement. Moreover, his brother, Ghulam Hussain, had already become an Ahmadi, the first in the family to do so. After a period of reflection and prayer, Muhammad Hussain resolved to travel to Qadian to visit the Ahmadi community and see for himself. There, Mirza Ghulam Ahmad's successor, the first *Khalifa* ('successor'), Mirza Nuruddin, was preaching in the mosque. After the death of Mirza Ghulam Ahmad in 1908, the charismatic Khalifa at the centre of the tightly knit Ahmadi community plays an important role. He is continually beseeched to advise on all matters. Vivid dreams are considered by Ahmadis to convey important messages, and the leader is requested to interpret them. At his first glimpse of the Khalifa, Muhammad Hussain hesitantly remained at the edge of the congregation, but was spotted in the crowd by Mirza Nuruddin, who asked him to come forward. Muhammad Hussain, impressed at having been so selected, soon made the pledge (*bai'at*)[3] to become an Ahmadi.

In his world travels, Abdus Salam became a surrogate Ahmadi figurehead, called upon to sign books, shake hands, pose for photographs, bless babies, and give career advice, all of which he dispensed with 'unflappable reserve'[4]. But the widely misunderstood Ahmadi sect has never been comfortable in Pakistan: when it was tolerated, Salam was successful, but most frequently Ahmadis have been the target of persecution. Even with the added buoyancy of a Nobel Prize, Salam was jettisoned by his country and sank into national obscurity.

Even during the lifetime of Mirza Ghulam Ahmad, Islam in the subcontinent had begun to move towards an increased self-awareness that would overshadow the exuberant new sect. Like the first shift of snow that precipitates an avalanche, these movements were at first imperceptible, then slowly gathered momentum, and finally becoming irresistible.

After the Revolution/Mutiny of 1857, the Indian Muslim community, about a quarter of the population, was enfeebled and embittered. Their traditional figurehead, the Mughal Emperor, had been removed from the scene and replaced as symbolic ruler by the distant figure of Britain's Queen Victoria. Their traditional popular roles as minor officials had weakened as British control grew, and as Persian, the language of their literature, was replaced as an official language by English. In this confusion, far-sighted men saw the need for a new focus of awareness and pride.

Syed Ahmed Khan, born in 1817, came from an aristocratic Mughal family and was among the first Indians to study in Britain. His maternal grandfather had been a distinguished mathematician. After serving in India's British-run judicial service, he went on to become a member of the Viceroy's ruling Legislative Council. Syed Ahmed Khan had witnessed what the British had done in Delhi after 1857. As well as improving their traditional precarious equilibrium with the Indian Hindu majority, he maintained that Muslims in India should come to terms with Western ideas. If the British were ever to leave India, who would safeguard the interests of the Muslim community? Throughout his life, Syed Ahmed Khan strived for Anglo-Muslim understanding. In 1875 he founded the Muslim Anglo-Oriental College in Aligarh, later to become the Aligarh Muslim University, where generations of Indian Muslim rulers and administrators would emerge to serve their community and later a new nation. (Aligarh, between Delhi and Agra, is now in India.) Although this university only provided a minority of the total graduate output, it was highly visible because of its Muslim commitment, and became an integral part of the network of institutes of higher learning established by the British in the late nineteenth century.

The educated class that emerged from these universities and colleges would go on to challenge colonial authority, but initially created a visible target for orthodox Sunni Muslims. The frustrated reformer retorted 'You may call me an infidel, a *kafir*, but allow me to educate the nation's youth for the future just as you allow a non-Muslim mason to build a mosque'[5]. Syed Ahmed Khan made a brave attempt to reconcile the Holy Qur'an with modern science. Fundamentalists found (and still find) it easy to reject much of the body of modern science, as it appears to run counter to the Holy Book. Syed Ahmed Khan pointed out that if the Holy Qur'an is indeed the word of God, and if the findings of modern science can easily be shown to be correct, any such contradiction must be superficial rather than real. But these suggestions fell on deaf ears, and Syed Ahmed Khan's stature as a Muslim nationalist far outweighs his reputation as a philosopher.

With an eye to future political involvement, the Indian Congress Party emerged under British patronage in 1885 as a body that could speak for all of India. Although it included Muslim representation, some of it quite heavyweight, Muslims felt it required a counterbalance, and in 1888 under the leadership of Syed Ahmed Khan formed a Joint Committee of the Friends of India as an opposition group.

In the Crimean War of 1853–6, European powers had sided with Muslim Ottomans against Russia. However, European powers keen to extend their empires soon began to look greedily at the Western parts of the Ottoman Empire. Soon, much of North Africa came under European control. Although the Ottoman Empire was only a fraction of the total world Muslim population, this dismemberment was a blow to their pride. This erosion accelerated in the twentieth century, with the Balkan Wars of 1912–13, in which the Ottomans lost almost all of their territory in Europe.

In London, the official stance towards the Indian Muslim population still remembered the trauma of 1857. However on the ground, enlightened Viceroys such as Mayo and Dufferin realized that something had to be done, and looked to improve the status of Indian Muslims, offering new opportunities to improve their status. This message was relayed by Syed Ahmed Khan and others, urging Muslims to use these new opportunities and thwart the traditional Hindu stress on examination success. The Ahmadi movement, with its pro-British stance, emerged during this time.

Although concentrated in eastern Bengal, the Punjab, Sind and Kashmir, Muslims were nevertheless scattered across India, and first-past-the-post elections would handicap Muslim parliamentary representation, despite a Muslim presence inside the Congress Party. As a harbinger of what was to follow, the province of Bengal had been split by Viceroy Curzon in 1901 for administrative convenience, without regard to its human implications. The decision was revoked ten years later, but it changed the mood of India. It galvanized the Joint Committee of the Friends of India, which in 1905 evolved into the All-India Muslim League. At first overshadowed by the Congress Party, the League slowly emerged as the Muslim voice of the subcontinent. A strong contingent was made up of lawyers who could build a legal framework to safeguard Muslim interests. One was the urbane and sophisticated Muhammad Ali Jinnah. Born in Karachi in 1876, he went on to qualify as a barrister in London, graduating to the Viceroy's Legislative Council and becoming president of the Muslim League. A pact signed in Lucknow in 1916 looked to assure peaceful coexistence between the Muslim League and the Congress Party, but the sudden appearance on the scene in 1915 of the mercurial Mohandas Gandhi and his unconventional tactics upset the fragile equilibrium. Outmanoeuvred, Jinnah quit India for London, leaving the Muslim League leaderless at a vital time.

In the First World War, Indian Muslim soldiers had to face a dilemma: called to fight for their British masters, they had to combat the forces of the Ottoman Empire. Indian Muslims remembered the proud Mughal tradition, and with the Mughals removed, this collective nostalgia had been reflected towards Constantinople. Indian troops did outstanding duty to their colonial masters[6], but the outcome, the ultimate defeat and final dismemberment of the Ottoman Empire, its temporal sultan deposed, was nevertheless a shock to Muslim pride. Like Indian Muslims almost a hundred years before, Muslims who had lived under Ottoman rule in the Middle East now found themselves leaderless. They were also the subjects of clumsy new French and British colonial administrations, and the effects of this seism continue to be seen.

For a time the Muslim world acknowledged a separate caliph on a precarious seat in a Constantinople rocked by civil war, and in 1924 two influential Indian Muslims – the Aga Khan of the Ismaili sect, and

Amir Ali – asked the Turkish government to safeguard the vestigial caliphate. For Turkey's implacable leader, Kemal Atatürk, this perceived foreign meddling in the affairs of his country was the last straw, and the caliphate was abolished. Abdülmecit and other surviving members of the Ottoman dynasty were banished. Asia Minor was a long way from India, but these events, underlined by some adroit moves by Gandhi, underlined the Indian Muslims' sense of futility and hopelessness. In the Punjab, there were riots in Multan and Lahore. Despite this insecurity, Punjabi Muslims benefited during this period from improved educational opportunities, the number of pupils increasing from 242 000 in 1921 to 543 000 in 1926[7].

Indian Muslims in general and those from the Punjab in particular were given fresh hope in 1930 when the European-educated poet Muhammad Iqbal, who had become a charismatic President of the Muslim League, made a loud rallying cry from a platform at Allahabad: 'I would like to see the Punjab, North-West Frontier Province, Sind and Baluchistan amalgamated into a single state. Self-government within or without the British Empire, the formation of a consolidated North-West Indian Muslim state appears to me to be the final destiny of the Muslims... of North-West India.' The idea motivated Rehmat Ali, a Punjabi Muslim student in England, to invent the acronym 'Pakistan' – Punjab, Afghans (Pathans), Kashmir and Sind, all areas or communities mentioned or implied by Iqbal, compounded with 'stan', the Persian suffix meaning 'country' – giving the word that in Urdu means 'land of the pure'. In 1934, Muhammad Ali Jinnah returned from his self-imposed exile in London and took back the reins of the Muslim League. Abandoning his earlier policy of reconciliation to live alongside Hindus and of building some form of Hindu–Muslim federation, Jinnah now looked to a nation with 'distinctive culture and civilization, language and literature,... history and tradition, aptitude and ambitions'[8]. Jinnah was to emerge as the Qaid-i-Azam, the Great Leader, a heavyweight contender to Gandhi and the Congress Party.

But attention was soon diverted by the Second World War. After the fall of its South-East Asian territories to the Japanese, the British, anxious to retain Indian goodwill, offered new independence incentives. With the war in Europe over in May 1945, Britain looked to use India as a springboard for continued attacks against the Japanese in South-East Asia. But the atomic bombs on Hiroshima and Nagasaki precipitated the surrender of Japan, and India and its army were no longer a part of Britain's military strategy. In Britain, Winston Churchill's wartime regime was replaced by a Socialist government voted in by an impoverished electorate tired of war. The new government reflected the population's goal of rapid demobilization, and was not prepared to make further efforts to maintain a strong British presence in India. British plans to leave India suddenly accelerated.

In the first elections since 1937, the Congress party was dominant overall, but the Muslim League, now solidly behind the creation of Pakistan, showed its strength in the Muslim electorate. A high-level mission sent by London in 1946 searched for a way of ensuring a united India that would safeguard Muslim interests. It failed. Political negotiation gave way to direct action, and slowly India began to slide towards civil war. Riots broke out in Calcutta, the worst ever seen in Britain's several hundred years of colonial history. The unrest soon spread to the Punjab. In Multan, several thousand lost their lives and a curfew was imposed. It was against this turbulent background that the young Abdus Salam prepared to leave his home and begin studies at Cambridge University.

In February 1947, in the middle of a post-war glacial winter with a nationwide shortage of fuel, with a major problem on its hands in its Palestine mandate, Britain's harassed government announced that power would be handed over in India no later than the following year, and that Lord Louis Mountbatten would take over as Viceroy for the twilight of British rule. He quickly added new impetus to the already accelerated withdrawal plan. Separate nations, India and Pakistan, would come into existence, and individual states and provinces should decide their future allegiance by legislative vote. But the large provinces of the Punjab and Bengal, with roughly equal Hindu and Muslim populations, would be split by a special British boundary commission. Charting such demarcation lines is a delicate business, but the hasty work of this commission can be seen as heavy-handed, with an eastern Punjab under Indian rule giving ready access to Kashmir to the north, and exercising control over the headwaters of the Punjab irrigation system. The Pakistan that finally emerged was smaller than its pioneers had envisaged, and voters had been led to believe. An embittered Jinnah, knowing that initially he would have no army to call on, nodded his approval to Mountbatten.

As sometimes happens with twins, the two children of British Mother India bore little mutual resemblance. In such a traumatic birth, they both suffered and were to bear scars. As midnight of 14–15 August 1947 struck, speeches were made, bands played and new flags hoisted. The results of the boundary commission decision were only announced several days later, a spark that ignited the inflammable mix of Hindus, Muslims and Sikhs in the Punjab, pitching the divided province into chaos. Sporadic warfare became mass terror and slaughter. Refugee convoys and trains moving in either direction were attacked, and even fought each other. Delhi was swamped by Hindu refugees and the Pakistani Punjab by Muslim fugitives. Millions uprooted themselves – some half a million died.[9] In an attempt to calm the storm, Gandhi returned to the political limelight, only to be shot by a Hindu fanatic. India was stunned.

In Multan, where Salam's family lived, several thousand were killed, and schools were closed for months. His family relate how Punjabi Hindus preparing to flee were selling their livestock at knockdown prices. Salam's father, Muhammad Hussain, was offered such a cow for 20 rupees, but altruistically bargained backwards and offered 40 instead, knowing it was worth 80 and that he still had a bargain. To aid incoming Muslim refugees, Pakistan offered subsidized plots of land. As the home of Salam's mother's family was now in the Indian Punjab, they qualified for a grant, but Ahmadi philanthropy instructed them not to benefit from such largesse, and to leave the land for others.

Pakistan came into being in 1947 as a homeland democracy where Muslims could live freely in accord with their religion and its traditions. With few role models to emulate, it was a political experiment. So much energy had been channelled by Jinnah and others into conceiving nationhood and attending its birth pangs that little forethought had gone into what would happen afterwards. In addition, it was handicapped by its geography. The Indus plain that had given the Pakistan acronym had been uneasily yoked together with East Bengal, 2000 kilometres away. In such a heterogeneous mix, one immediate problem was an official language. In Bengal people spoke Bengali, but the Indus plain had its own palette of tongues. Educated refugees from the provinces of Uttar Pradesh and Andra Pradesh pushed for the adoption of classical Urdu, despite the fact that few Pakistanis knew or even understood it[10].

The people of Pakistan worshipped Jinnah, but he was already infirm and died in September 1948, leaving a vacuum difficult to fill. In the ensuing volatility, Prime Minister Liaquat Ali Khan was assassinated in Rawalpindi in 1951, a macabre reflection of Gandhi's murder. Traumatized, Pakistan tried to tread the tightrope of its avowed Muslim cause without penalizing remaining minority elements. But these attempts only drew more attention to such minorities. Jogendra Nath Mandal, the Hindu Labour Minister, who had been a close associate of Jinnah, quit and went to India. Increased religious pressure in the drafting of the nation's constitution led to proposals, such as reserved seats for women in the National Assembly, being vetoed as un-Islamic.[11]

Another ready target was the Ahmadi movement. After its foundation in the late nineteenth century, the community had thrived, setting a tradition of enterprise and study, profiting from the British infrastructure. However, after Hindus and Sikhs had fled for India, the Ahmadis became more visible. The murder of an Ahmadi army officer in 1948 underlined their new plight, and they sought government protection, which only increased their visibility. Pakistan Foreign Minister Muhammad Zafrullah Khan, as the country's most visible Ahmadi, had always been a target for criticism, and Jinnah had resisted demands to remove him from the cabinet. After Jinnah's death, the clamour to remove Zafrullah Khan revived. As anti-Ahmadi tension in the Punjab mounted in the early 1950s, the government in Karachi had much else to worry about, and turned a blind eye to the worsening situation, judging that it was a matter for the Punjab administration, where the 1951 elections had opened the door to vociferous religious elements. For the second time in four years, the Punjab was torn by riots and bloodshed, but this time the target was Ahmadis rather than Hindus or Sikhs. This was the situation that enmeshed Abdus Salam soon after he returned from Cambridge and Princeton in 1951.

The government eventually imposed martial law in the Punjab, but the infant nation had been rocked by a storm that would not be easy to forget. Plagued by indecision and blindfolded by prejudice, the country groped its way forward. A strong leader was needed to sweep away the accumulated debris of partition and set the country on a firm course. The spark came in October 1958 in Baluchistan, a region that had always been reluctant to join the Pakistan family. The Khan of Kalat, south of Quetta, seized the Miri Fort and hoisted his own flag in place of the national standard. The Pakistan army soon moved in and deposed the rebel leader. It would have been just another colourful incident in Pakistan's mountainous frontier region, except that the next day the army moved into Karachi, Lahore and other main towns. Commander-in-Chief of Pakistan's Army was the forceful figure of Ayub Khan. Educated at the Aligarh Muslim University in British India and at the prestigious Sandhurst officers' training college in Britain, Ayub had commanded an infantry battalion in the Second World War and served in the Punjab Boundary Force during the 1947 partition.

After a face-off with President Iskandar Mirza, Ayub Khan became the new President of a country on which he imposed martial law, and flamboyantly awarded himself the title of Field Marshal. Anti-Ahmadi demonstrations were immediately crushed, and the followers of Mirza Ghulam Ahmad emerged timidly from hiding, once more free to go about their business.

However, Ayub Khan's objectives were wider than to gain personal power, and his enlightened thinking introduced an autocratic political structure together with rigid measures to stabilize Pakistan's economy. Until 1958, Pakistan's capital had been the port of Karachi in Sindh in the far south. The Lahore riots of 1953 had displayed the problems of having such a remote centre of government, and the continued concentration of investment and development in the capital city threatened to upset the country economically. In his characteristically impassioned style, the visionary Ayub Khan moved to create a totally new capital immediately north of Rawalpindi, near the Margalla hills and the historical site of Taxila. In 1967, Islamabad, gridded by green belts and parks, became the country's symbolic new capital. But Pakistan had many pressing problems to face, and had dallied for too long. A key component in Ayub's perception of the nation's future was a new scientific and technological thrust, and Abdus Salam in London would be the man to oversee it.

REFERENCES

1. www.alhafeez.org/rashid/ludhianvi/abdussalam.html
   Bashir, A., *The Ahmadi movement, the British-Zionist connection*, Pakistan, 1994
2. See ref 1
3. http://www.alislam.org
4. Crease, R., Mann, C., *The second creation, makers of the revolution in twentieth-century physics* (New York, Macmillan, 1986)
5. Salam, A., (ed. Dalafi, H. R., Hassan, M. H. A.) *Renaissance of science in Muslim Countries*, (Singapore, World Scientific, 1994)

6. During the First World War, Mahmud al-Hasan Sheikh al-Islam (1851–1920) went to the Hejaz to make contact with Turkish military leaders, but was arrested by the forces of the Arab Revolt under Sherif Hussain of Mecca and handed to the British, who interned him in Malta.
7. Hardy, P., *The muslims of British India* (Cambridge, CUP, 1972)
8. Rajput, A. B., *Muslim League, yesterday and today* (Lahore, Muhammad Ashraf, 1948)
9. Spear, P., *History of India* Vol. 2 (London, Penguin, 1990) Khosla, G. D., *Stern reckoning*, (1950, New Delhi)
10. Urdu was less of an obstacle for those who knew Punjabi.
11. Ziring, L, *Pakistan and the 20th century* (Karachi, OUP 1997)

4

A mathematical childhood

Were it not for its five rivers, the Punjab would be desert. But as the Nile does for Egypt, these rivers paint the desert with fringes of green. Over the centuries, far-sighted regimes have built irrigation systems around these rivers, extending the green and pushing back the barrenness. In the early nineteenth century, a major project by the British began by repairing the vestiges of Mughal engineering, so buffering local agriculture against the annual lottery of the monsoon. As the twentieth century began, people toiled in fields of subsistence cereals. Apart from the irrigation schemes, technological advances hardly touched them, and life continued much as it had done in the days of Heer Ranja. At the end of their day's work, the men would gather under the trees to talk, drink tea and perhaps listen to a recitation of Waris Shah's poetry. Few of them could read, and they would wait for a literate man to tell them what was printed in the newspapers that arrived on the train from Lahore. A mighty conflict had engulfed Europe. The fighting was far away, but Indian soldiers were dying in the mud of France. The powerful Indian Army was also fighting for its British masters against Muslim Ottoman troops in Mesopotamia. As the conflict subsided, the daily news turned to the wider Punjab, where a multiethnic coalition was striving to safeguard the province's agriculture. Elsewhere in India, Mohandas Gandhi and the Congress Party were trying to loosen the assimilated shackles of British colonial rule; and Indian Muslims rallied to support the ailing Islamic Caliphate in Constantinople. Later, newsreaders related how Mustafa Kemal Atatürk had abolished this caliphate, changing the face of modern Turkey and leaving Islam without its traditional figurehead. In Jhang, Abdus Salam's father, Chaudry Muhammad Hussain, read these newspapers, but did not sit under the trees to discuss them with the other men. After his long working day, he returned home to his family to pray, to speak with his children, and to prepare himself for the next working day.

An ancestor of Muhammad Hussain, Hazrat Sayed Budhan, originally a Hindu princeling from Rajputana, had converted to Islam in the twelfth century[1]. The lineage has a tradition of such honorific titles. Muslims have a keen awareness of pedigree. While in the West this knowledge normally spans just the few generations of living memory, a Muslim family can recite eleven generations of forebears as easily as Europeans recall the lineup of their favourite football team. In 1911, Muhammad Hussain had gone to study at Islamiyya College in Lahore, but did not formally graduate[2]. His brother, Chaudry Ghulam Hussain, older by 14 years, had excelled in his school examinations and graduated from Forman Christian College in 1899. Entering the provincial government's education service as a teacher, Ghulam Hussain went on to become District Inspector of Schools. He was also the first in the family to embrace the new Ahmadi faith.

It was a major sacrifice for the family not to have strong boys available for manual work, but this was the price of a tradition of modest scholarship. Then, as now, higher education does not always mesh with the demands of everyday life, and when Muhammad Hussain returned to his home town from Lahore, no clear career path lay open. Resolutely, he took the initiative and asked if the local Jhang school needed an extra teacher, and was temporarily put in charge of a problem class of 'naughty boys'[3]. There, he initially earned 19 rupees a month, a fraction of the salary of a regular public service job (at the time one pound sterling was worth about seven rupees). After two years of probation, he progressed to a proper position, teaching English and mathematics to generations of Jhang boys. Gul Mohammad, the father of Ghulam and Mohammad Hussain, had been a skilled *hakim*, or healer, dispensing herbal medicines, and his schoolteacher sons continued this tradition. Muhammad Hussain had a small dispensary in his house where people with minor ailments could get advice, traditional medicines and a cup of tea.

Soon after his marriage in 1922, Muhammad Hussain gave thanks to God for his new daughter, Masooda Begum, and looked forward to a larger family. Instead, within six weeks he was a widower. With no secure job and with his family life now in ruins, his personal ambitions lay shattered. Still only 31, and with a child to look after, he needed a new wife. The Imam of his community, in a traditional role of matchmaker, located Hajira Begum, daughter of Hafiz Nabi Bakash, a *patwari*, or provincial government tax officer, in the village of Santokdas, in the

Sahiwal district, 60 miles south of Jhang. It was a family with a strong religious commitment: Hajira Begum's brother spent twenty years as a Muslim missionary in West Africa. Muhammad Hussain and Hajira Begum were married in May 1925, and his ambition was sublimated into his future offspring: soon came his incandescent vision of the coming of a son called Abdus Salam, and he eagerly began to make plans. To ensure that the vision would become reality, the most important thing was to offer fervent daily prayers: Muhammad Hussain must have spent a significant fraction of his adult life praying for the success of his son. As was then the custom, when her time approached, Hajira Begum returned to her family home in Sahiwal so that her mother and sisters could attend the birth of Abdus Salam on 29 January 1926. After the customary confinement period of forty days, mother and child returned to her husband's two-roomed mud house in Jhang.

The baby, with black, curly hair and wide eyes, was doted on by his parents, especially his father, who would parade the boy around town on his bicycle. Even when brothers and a sister arrived in rapid succession, Abdus Salam was given his own corner of the tiny house. His younger sister, Hamida Begum, helped with household chores and later became his personal handmaid, cleaning and folding his clothes, and following local custom, smearing black kohl around his eyes. The infant Abdus Salam was a bonny child, and at the age of two was judged the town's most healthy baby, the first of many awards in his lifetime. As a schoolteacher, Muhammad Hussain knew that his son needed intellectual stimulation, and continually recited stories and poems to exercise memory and stimulate expression. His mother taught Abdus Salam to read the Holy Qur'an and write in Arabic script, where the name of Allah introduced the alphabet. When the boy began to read books, his father would ask him to summarize what he had read, correcting his use of words and his spoken delivery. The Ahmadis place immense importance on the authority of their religious leaders, continually soliciting their opinion on all matters, and requesting them to echo and reinforce personal prayers. Soon after Abdus Salam's birth, his father asked a visiting community representative, *Maulvi* Ghulam Rasool Rajiki, respected as a pious and religious man, to pray for the young boy. The Mauvli foretold that the boy would one day speak so loudly that the world would listen. Muhammad Hussain was impressed. This was now the second time that a remarkable career had been prophesied for his son.

One of Abdus Salam's earliest memories was of learning the multiplication table for the number fifty. It is not a particularly difficult table to master, but on the other hand an accomplishment when most children are still learning to count to ten, let alone fifty. As a boy, he was intrigued by engines – railway locomotives, motorcycles, flour mills, cotton gins. At play, he would make model irrigation schemes in the dirt and watch with glee as he opened his carefully constructed dam and the channels would flood with water. Jhang's overlord was Inyat Khan Sayal, a cruel, illiterate man who got his pleasure by arranging fights between bears and dogs. While ploughing a field near Salam's house, a peasant farmer unearthed a cache of silver coins, which he duly brought to the local ruler. Deeming them worthless, Inyat Khan Sayal threw them to children in the street, but the young Salam was too small to fight his way through the crush.

Although Muhammad Hussain was now a career schoolteacher, life was still hard. His initial monthly salary of 19 rupees did not stretch to meat every day for a family with nine children. New clothes were a remote luxury. Seeking extra space, in 1931 Muhammad Hussain built a modestly larger house, with two rooms upstairs and two rooms down. Next to the house, there was enough land for chickens, two goats, a cow and a buffalo. These animals quickly exhausted what grew on the land, and the family had to buy a daily load of green fodder, further eroding the meagre government salary. With no electric light, reading or study had to be outside, or crouched round an oil lamp. To avoid Abdus Salam interrupting his studies, the lamp was cleaned and fuelled by his brothers and sisters. Meals were simple, mainly chapati bread, lentils and vegetables, occasionally a small piece of meat. One day Abdus Salam was so absorbed in thought that he dropped his morsel of meat on the floor, where it was seized by one of the domestic chickens. There was no second helping.

Every day Muhammad Hussain continued to recite verses, tales, and the Holy Qur'an to his son. It was from his father that Abdus Salam learned to respect and love Islam. The word of God as revealed to Muhammad has its own melodic beauty that underlines the powerful messages and allegories. The Holy Qur'an uses the dialect of Arabic that was used in Mecca and Medina in the seventh century and its holy message cannot be edited. Its preservation guarantees the purity of classical Arabic. (In comparison, the continual evolution of English means that the language of Shakespeare, a thousand years younger than the Holy

Qur'an, although understandable, is already unfamiliar, but the Anglo-Saxon of the seventh century is effectively a foreign language, decipherable only to scholars.) The Holy Qur'an is read, but the prepared mind hears the traditional mesmerizing incantation. The Holy Qur'an is not a narrative, but returns again and again to its underlying themes. This later became a feature of Salam's writing in English.

At Jhang school, just five hundred yards away, boys normally began preparatory learning when they were six. Muhammad Hussain had judged that his son was ready for instruction at the age of three. It is one thing for a gifted child to be a precocious learner, but another to have to contend with bigger boys pushing and shoving. When Abdus Salam was finally admitted to primary school at the age of six, he was soon transferred to the middle school, alongside boys several years older. In these village schools, boys (girls did not yet go to school) sat on the floor and wrote on a slate. The 70 boys were taught by three teachers. Initial lessons were in their Punjabi mother tongue, but pupils soon learned Urdu. Rote learning in school was no problem for Abdus Salam, who looked forward every day to his evening study sessions with his father, learning more multiplication tables, and writing pieces for the children's section of the local newspaper. His primary school accomplishments included prizes for penmanship, map drawing and first aid[4].

When the time came to leave the primary school in 1938, Muhammad Hussain's plan was to send his son to the Central Model School for Muslim boys in distant Lahore. There, he judged, his able son would be able to forge ahead, and get a good grounding in English, which he knew was vital for any achievement other than as a religious scholar. On paper, Abdus Salam's candidacy looked watertight. Muhammad Hussain always wore a Punjabi turban, and so did his son. but his father made the mistake of telling his son to wear a red fez (*Rumi topi* – Roman hat) during their visit to Lahore. This distinctive headgear had been widely adopted across the Ottoman Empire and copied by Muslims further afield. The Muslim visionary Syed Ahmed Khan had popularized it in nineteenth century India. With no brim, it was good headwear for people whose foreheads must touch the ground each time they pray. However, in 1925, as part of his push to drag his country into line with twentieth-century Europe, the Turkish leader Mustafa Kemal Atatürk had ordered that Turks should abandon traditional dress and wear Western clothes. This included abandoning the traditional fez in favour of a hat or a cap with a brim. Atatürk set the trend by wearing a panama. Almost overnight, the fez became old-fashioned, but in 1938 this message had not yet reached Jhang. The headmaster of the Lahore Central Model School was not impressed by the hayseed boy, and thought it unwise to take on a new pupil, however gifted, who would be cruelly mocked by the street-smart town boys. Crestfallen, Muhammad Hussain took his son to Jhang's Government Intermediate College, where the majority of students were Hindu[5].

Nevertheless, Abdus Salam enjoyed his second school in Jhang. In his first exams, his 591 marks out of 700 placed him first in the district and fifth in the province, qualifying him for a two-rupee book prize. Soon, a scholarship of six rupees per month was a useful addition to the family budget. He was put in charge of the class library, and quickly made friends. Later, he became Editor of the college magazine, *Chenab*. On several occasions in later life he expressed his gratitude to have had such wise and affectionate teachers. Shaikh Ijaz Ahmad was his English teacher, Soofi Zia-ul Haq taught Arabic, and Khawaja Mirajud Din taught Persian. Mathematics and science were usually taught by Hindu and Sikh teachers[6]. The science teacher spoke of the gravitational force that kept the Earth, the Moon and the planets in their orbits, and Salam later recalled that Newton's name had penetrated even to a place like Jhang: 'Our teacher then went on to speak of magnetism, and showed us a magnet. Then he said 'Electricity! That is a force which does not live in Jhang, it lives only in the capital city of this province, Lahore.'[7] Electric lighting had not yet reached Jhang. And what of the nuclear force? 'That was a force that only lived in Europe. It did not live in India and we were not to worry about it.' Salam related, 'but I still remember that he was very keen to tell us about one more force – the capillary force, which according to him was a fundamental force of Nature. Most likely he was following the ideas of Avicenna (Ibn Sina), who was a physician as well as a physicist, and there was no force more important than one which makes the blood rise in the smaller capillaries.'[8]

As his command of English improved, Salam began experimenting with fancy words and phrases, without first checking their proper meaning or context, and peppering his texts with quotations. Despite his teacher's warning, he found this fun and stubbornly continued. When exam time came, he duly lost five points for each wrong word, with disastrous effects on his performance, and the teacher read out Salam's efforts to the entire school. Humiliated, Salam finally complied with the teacher's instructions. But he did not bear a grudge. Later he said 'I feel that it was the proper medicine administered to me. The net result of this shock therapy was that I stopped using difficult words altogether.'[9]

English was later to become Salam's main instrument of communication. In the British Raj, English had supplanted Persian as the language of administration, culture and commerce. Classical English literature was there for everyone to assimilate, but the language was taught formally by non-native speakers. In the days before recordings became widely accessible, this inevitably left its mark. The English used in India developed an identity of its own, technically so correct as to be almost pedantic, but often with unusual constructions, such as 'he is knowing the answer', and syntactical peculiarities like 'affectee' (someone affected) and 'allottee" (someone allotted). As British influence waned, archaic words like 'thrice', 'conveyance', 'brigand' or 'interloper' were nevertheless left stranded in common use. Modern Indian newspapers still tell of 'miscreants' or 'vagabonds' who 'abscond' after doing 'misdemeanours', but eventually get 'nabbed' by 'sleuths'. Bereavements are condoled and successes are felicitated. In a continual effort to be modern, the contemporary English of the subcontinent sometimes looks as though it has leapt directly from Kipling to the twenty-first century – 'He demanded that these phone chats should be made public'. The English of South Asia is sprinkled with local words such as 'lakh' (a hundred thousand) and 'crore' (ten million, or 100 lakh). On the other hand, the subcontinent has donated many words to modern English, such as *khaki* ('dust coloured'), *pyjama* ('leg garment'), *bungalow* ('country house'), *thug* ('thief' or 'rogue'), *coolie* ('porter/labourer'), and *hullabaloo* ('great noise'). Salam later paraded his English vocabulary skills to the full, describing failed students as 'broken reeds', a mutual agreement as a 'compact', and laboratory apparatus that was not user-friendly as 'recalcitrant'. His pedantic teachers and the grammatical rigour of Qur'anic Arabic made him keenly aware of the limitations of English, whose lack of inflection can sometimes lead to confusion. When he wrote of an uncomfortable night on a plane full of 'crying servicemen's children', he added quickly, 'that is, the children were crying, not the servicemen'[10].

Muhammad Hussain knew that Hindus considered formal education to be especially important, and having his son studying alongside keen Hindu students softened the disappointment of not getting a place at the Muslim school in Lahore. In 1940, after several years at

Jhang College, the challenge of the Matriculation and School Leaving Certificate loomed. Muhammad Hussain desperately wanted his son to score the highest marks. Students from other schools traditionally performed well, especially those from Hindu Sanatam Dharam and Arya. Muhammad Hussain detected that his son had spidery handwriting, not good if an examiner was tired or had poor eyesight, and that other weaknesses were practical science, geometry, and translation into Arabic. To remedy these perceived deficiencies, he tried to organize extra coaching for his son. The Principal of Jhang College was at first unco-operative, pointing out that his job was to boost the overall pass rate, not to ensure that one particular candidate would come first in the rankings, but then became more helpful[11]. Salam's literary efforts were coached by a local poet, Sher Afzal Jafri[12]. Translation into Arabic was supervised by a Muslim teacher, and practical science by a Sikh. These efforts were supplemented as usual by the family's earnest prayers.

After the exams, Muhammad Hussain had a vivid dream that culminated in him being awarded a cup of delicious syrup made from a *neem* tree. For him, this was an auspicious sign. When the results were due to be announced, Salam was waiting in his father's office. The newspapers from Lahore with the exam results were expected to arrive around lunchtime at Jhang station, but even before the train steamed in, telegrams of congratulation had started to arrive. Salam recalled cycling home in the early afternoon, when the news of his 'standing first' in the exam had already arrived. Passing through the city to reach his home, Hindu merchants who normally would have closed their shops in the afternoon heat stood outside to congratulate him. The local newspaper later carried a photograph of a startled-looking boy peering out through newly acquired round-framed glasses under a turban, the fez having been discarded (see Plate 6). His government scholarship increased to 20 rupees per month, which was supplemented by another 30 from an Ahmadi award instituted for the first time in 1939, for students classed in the top three of university examination lists.

On this performance, Abdus Salam could have moved to Lahore, this time to its Government College, affiliated to Punjab University, but fourteen is a tender age to be channelled into specialist study, and it was still not clear in which direction Abdus Salam's talents should be focused – on the one hand mathematics and science; or on the other language and literature, whether English, Urdu or Persian. Muhammad

Hussain sought the advice of a distinguished member of the Ahmadi community, Muhammad Zafrullah Khan. In 1940, Zafrullah Khan, educated at Cambridge and a London-trained barrister, was a member of the ruling Executive Council of the British Governor-General of India, and would soon be one of the judges of India's Supreme Court. Apart from the religious leaders of the movement, he was one of the most visible and therefore influential Ahmadis. Zafrullah Khan replied that he would pray for Abdus Salam, and offered three pieces of advice: firstly, the boy should look after his health, for this was the foundation for all achievement; secondly, all lessons should be prepared for beforehand and revised immediately afterwards; and thirdly, that the boy should broaden his outlook, especially through travel[13]. The paths of Zafrullah Khan and Abdus Salam were later to cross again: the advice written to Muhammad Hussain in 1940 was the harbinger of a major and lasting influence, which would eventually propel Abdus Salam onto the stage of the United Nations.

Abdus Salam continued at Jhang's Government Intermediate College for another two years, passing his FA (Faculty of Arts) exams in 1942 with 555 marks out of a possible 650, again 'standing first' in the whole of the Punjab. Salam's monthly scholarship increased to 30 rupees from the government and 45 from the Ahmadi fund. Zafrullah Khan's advice on how to learn had been valuable, but still driven by his 1925 vision, Muhammad Hussain was looking further ahead, to what career his son should embark on once his education was complete. He had the seeds of an idea, and corroborating advice came from Mirza Bashiruddin Mahmud Ahmed, the leader of the Ahmadi community.

As independence loomed, it was clear that the few Indians recruited by Britain into the Indian Civil Service (ICS) would ultimately play a major role in the future of their country. Making up only a quarter of the population of India, Muslims were naturally a minority in any national activity. Few Muslims achieved ICS status. Whatever the future of Indian Muslims, whether as an autonomous region inside an Indian federation, or as a separate country, they would need skilled administrators, and this was the career path that Muhammad Hussain charted for his son. It was an ambitious choice and a demanding one. Apart from religious studies, Indian Muslims were not prominent in higher education in the subcontinent. For entry into the Civil Service, Abdus Salam would have to study at a British university. Getting admitted was only half the problem: the other was to get a scholarship. Only the Indian aristocracy could pay their own way at British universities.

In 1942, Muhammad Hussain was promoted to the administrative offices of the Punjab Education Department in the town of Multan, the capital of the South Punjab and 100 miles from Jhang, overseeing teaching assignments and transfers in high schools all over the province. His monthly salary increased to a fairly generous 250 rupees. Multan, claimed to be the oldest surviving city in the subcontinent, was already old when visited by Alexander the Great in 325 BC. It had been briefly conquered by the Islamic army of Muhammed Bin Qasim in 712, and largely flattened in 1848, when the British laid siege there to Moolraj the Sikh. Later, Multan horsemen went with the British to Delhi in 1857. The town has a reputation for heat, dust, beggars and burial grounds.

With the whole family set to move, the sixteen-year-old Abdus Salam was ready to make his next step up the academic ladder and transfer to Government College, Lahore. The capital of the Punjab for a thousand years and rich in history, Lahore was a very different place from the dusty market town of Jhang, or even Multan. Already firmly established in the eleventh century on the banks of the Ravi River under the rule of the Ghaznavid dynasty, Lahore reached its apogee five hundred years later when Akbar ('The Great'), the third Mughal Emperor, made it his capital from 1584–98, building the impressive central fort and its enclosing red-brick wall. Subsequent emperors extended the fortress, adding more exotic palaces and the beautiful Shalimar gardens, with their canals and sparkling fountains. Akhbar's son Jahangir has his mausoleum in the city. Aurangzeb, the last great Mughal emperor, built the impressive Badshahi mosque and its huge courtyard.

Later, the Sikhs, India's youngest religion, were eager to establish a homeland. At the beginning of the nineteenth century, their leader Ranjit Singh declared himself ruler of the Punjab and made Lahore his capital, relegating the proud Badshahi mosque to the role of a powder magazine. The stately palaces were pitilessly plundered to rebuild the Sikh Golden Mosque at Amritsar after it had been sacked by Afghan invaders. The arrival of the British brought fresh prosperity and a new influence, adding splendidly pompous buildings in their incongruous Mughal–Gothic style. To the north of the Old City is the central railway station, built by the British in the mid-nineteenth century as part of their grand plan for the Indian railway network, but conceived also as a fortress. During the partition of India and Pakistan in 1947, the fate of Lahore was undecided until the last moment, when the border between the two nations finally fell 25 kilometres to the east. As soon as it was clear that Lahore had been judged to be in Pakistan, the station became the focus of hordes of refugees fleeing in either direction. The dense mass of humanity was an easy target for rampaging mobs. Corpses piled up in and around the station, and trains arriving from Delhi would pull in with many of their passengers dead[14].

Government colleges all over India offered comprehensive education, with sport and recreational activities as well as formal lessons, providing good training grounds for careers in commerce and administration. Lahore's Government College had been established in 1864 as part of the move to set up a nationwide network of higher education. Initially affiliated to Calcutta University and housed in a royal palace, it moved into its purpose-built accommodation in the city centre 1871, with Gottlieb Wilhelm Leitner, professor of Arabic and of Muslim Law at King's College in London, as its principal. The gothic structure that dominates the campus south of the Old City looks from afar like a transplanted British parish church built from a kit of architectural spare parts, embellished with wide verandahs and high ceilings. Like many city buildings of nineteenth-century India, the British influence looks slightly bizarre against a backdrop of tall palm trees. The impressive Main Hall (now called Abdus Salam Hall) with its distinctive clock tower has a central, soaring nave surrounded by four double-storied aisles. Later additions tried to mimic the style of the original building. Government College became a university in its own right in 2002.

At Lahore's Government College, Abdus Salam was a boarder at New Hostel, receiving grants totalling 60 rupees a month from the local government and from the Ahmadi movement. Living away from home for the first time, he discovered chess and spent many hours pondering over moves to beat Hindu and Sikh players, before being reprimanded in his father's letters that he was wasting valuable study time. Salam also encountered a real mathematician for the first time. Sarvadaman Chowla's father also had been professor of mathematics at Lahore. After his master's degree at Government College in Lahore, Sarvadaman Chowla went to Cambridge to do research under John Littlewood, who, with Godfrey Hardy, was one of the leading British mathematicians in the early twentieth century, and who had guided Ramanujan. Returning to India after his doctorate in 1931, Chowla became in turn professor of mathematics at St. Stephen's College, Delhi; Benares Hindu University; Andhra University in Waltair; and eventually Government College, Lahore, where he was Head of Mathematics from 1936 to 1947. Chowla was a dedicated and prolific worker, collaborating with many other specialists and producing more than 300 papers. Although having virtually no interest outside mathematics, he was a lively and pleasant man, and is commemorated by mathematicians in the Bruck–Chowla–Ryser theorem, the Ankeny–Artin–Chowla theorem, the Chowla–Mordell theorem and the Chowla–Selberg formula. Most of this work was done after he, as a Hindu, left Lahore in haste in 1947, eventually proceeding to a series of senior posts in US universities.

As well as being a distinguished mathematician, Chowla was an accomplished teacher, making the subject come alive with illuminating examples and elegant proofs. Abdus Salam had the fortune to fall under his influence at Lahore. Chowla frequently ended his classes by setting profound homework questions. One focused on a problem attacked by Ramanujan a quarter of a century before: the solution of three simultaneous quadratic equations:

$$x^2 = a + y\,;\, y^2 = a + z\,;\, z^2 = a + x$$

The next logical step beyond Ramanujan was to attack the solution of four such simultaneous equations:

$$x^2 = a + y\,;\, y^2 = a + z\,;\, z^2 = a + u\,;\, u^2 = a + x$$

This was the problem Professor Chowla set his Lahore class of seventeen-year olds. A few days later, Abdus Salam triumphantly returned with a solution. It began 'Suppose x, y, z and u are the roots of a biquadratic (quartic) equation.'

Every scientific advance, no matter how small, transforms dissonance into harmony, disorder into symmetry. To do this requires an act of creation, a spark of intuition, which jumps between two things previously unrelated. To make such a spark, there is no button marked 'solve'. The hard shell of the problem must be held up to the light and viewed from other angles, until a previously invisible crack becomes visible – an unsuspected point of entry. There it was in the young Salam's solution. The 'suppose', which turns the problem round and sees it as part of a wider, more symmetrical, view. Applying this new lever, the hard shell of the problem fell apart effortlessly. After explaining the technique in just over a page of elegant algebra, Salam proudly concluded 'By employing the same methods, we can solve the (simpler) system of equations

$$x^2 = a + y \, ; y^2 = a + z \, ; z^2 = a + x$$

much more rapidly than Ramanujan did. His is a very laborious method.' Where the wonder of Madras had simply cranked a handle, the young Salam had searched for subtlety and elegance. The jubilant Chowla sent the solution to be published in the March–June 1943 issue of *The Maths Student*, a quarterly newsletter for Indian mathematical enthusiasts and wannabe Ramanujans[15].

As well as mathematics, at Lahore Salam initially continued with English and Urdu. He enjoyed classical Urdu poetry, fitting for a boy who had grown up in the same setting as Heer Ranja, and was steeped in the lore of the Urdu bard Ghalib. An article on Ghalib by the young Salam appeared in the Urdu review *Abdi Duniya* (Literary World). In English, Salam enjoyed the rapier wit of Oscar Wilde, and heavier stuff, such as T.E. Lawrence's 'The Seven Pillars of Wisdom', with its theme of Arab renaissance. Salam also fell in love with Urmila, the beautiful elder daughter of G. D. Sondhi, the College Principal.

To avoid such distractions, before examinations, Salam would lock himself in his room and get someone to pass sustenance through the window. Such dedication to studies made him overlook the statutory physical training session, always his Achilles' heel. When he was fined one anna (1/6 of a rupee), he glibly talked his way out and showed a clean college report sheet. As he had done in 1940 and 1942, his 1944 exam results established a new Punjab record. For good measure, this time Salam also sat the additional papers for the BA Honours English degree, and broke another record. Scholarship money increased to 120 rupees per month, equally split between government and Ahmadi funds. This brought a dilemma, as the results qualified Salam to continue for a master's degree in English or in mathematics. Confused, Salam sought the advice of the spiritual leader of the Ahmadi community. Mirza Ahmed recalled Salam's father's objective of the Indian Civil Service, then highly prized as the top career path for local talent. Recruitment was on hold during the Second World War, but, with independence on the horizon, able administrators would be needed to take over from the British and steer an independent nation towards maturity. To pass the demanding Civil Service entrance requirements when the moment presented, Salam opted to continue with mathematics, ably guided by such intellects as Chowla. Literature and general knowledge he could absorb on his own.

At Lahore, Salam's close colleague was Ram Prakash Bambah, whose career would go on to overlap with Salam's again at Cambridge. After several posts at US universities, Bambah became Vice-Chancellor of Panjab University, Chandigarh, India, and took the chair of the Indian Mathematical Society in 1969. Indian mathematics is coloured by the memory of Ramanujan, and Bambah won several of these memorial awards. Bambah has recalled the close camaraderie of those Lahore days when a fellow student had to be rushed to hospital with appendicitis[16]. As is the custom in the subcontinent, hospital nursing is provided by family and friends, and Salam spent 48 sleepless hours at the hospital. He also dispensed plenty of laughter, another good medicine.

As well as his academic achievements, Salam became President of the College Union, and Editor-in-Chief of the college magazine, *Ravi* (named for the river that flows through the city). Because of the plurality of local languages, the magazine had separate editors for Urdu, Hindi (at one time Ram Prakash Bambah) and Punjabi. A florid June 1945 Ravi short story 'The White Arm' by Salam, then the magazine's joint editor, displays a riot of vocabulary in a style amalgamated from Kipling, Rider Haggard and Conan Doyle.

In 1946, Salam topped the results list for the MA mathematics results at Government College with 573 marks out of 600, and stepped back to consider the next steps in his relentless climb upwards. But the wartime signposts were still awry. (His academic achievements had already opened the possibility of an engineering apprenticeship in the Indian railways, but this had quickly been discounted by the family as being incompatible with their ambitions, and by the Indian railways, when they found that Abdus Salam wore thick glasses.)

(In a staggering coincidence, another student at nearby Punjab University at this time was Har Gobind Khorana, born in a Hindu family in the tiny Punjabi village of Raipur, where his father was the *patwari* (village taxation clerk). As the only literate man in Raipur, the *patwari* had schooled his son, who progressed to the Multan High School and then to Punjab University in Lahore, where he earned a BSc in chemistry in 1943 and an MSc in 1945. A Government of India Fellowship allowed him to go to Liverpool University in Britain, where he earned a PhD in 1948, and moved through a series of postdoctoral positions. From 1950–52 he worked at Cambridge, where he briefly, and again unknowingly, overlapped with Abdus Salam. Khorana went on to share the 1968 Nobel Prize for Physiology and Medicine with Robert W. Holley and Marshall Warren Nirenberg for their work on the interpretation of the genetic code and its function in protein synthesis. Korana and Salam never met.)

The Second World War was heard in India at first only as a distant rumble. Then in December 1941, Japanese forces launched a surprise multipronged assault outwards across the Pacific. As well as the 7 December air raid that caught the US Fleet unawares in Pearl Harbour, within a few hours Japanese forces also invaded Malaya and Hong Kong, important easternmost outposts of the British Empire, and the Philippines, which until 1935 had been under US control. By March 1942, the Japanese had advanced northwards and taken the Burmese capital of Rangoon, overcoming Indian troops under British command. By May, the Japanese had crossed into India at Imphal and Kohima, which were to be the scene of fierce battles over the next two years. The humiliating fall of Singapore, with its disastrous loss of face for the British, suddenly highlighted the fragility of European colonial rule. Local populations everywhere took note. A Japanese invasion of India would be a further Asian disaster for the British, but could hasten the arrival of Indian independence.

In the Second World War, the British Indian army was more than two million strong, fighting in North Africa and in Europe as well as defending eastern frontiers against the push from Japan. While the smoke and thunder of the war raged, the exact future of the Indian Civil Service, like that of the country itself, was unclear, and recruitment was put on hold. During the war, the government of Sargodha district had dutifully collected 150 000 rupees to support the war effort, but at war's end, some of this still lay unspent in the Punjab administration's coffers. Khizar Hayat Tiwana, who had masterminded the scheme, became Punjab's representative in the Indian Congress Party and proposed using this money for scholarships to enable sons of poor farmers to study overseas. Here was an opportunity for Salam, but his father was a school inspector, not a farmer. Muhammad Hussain told his elder brother of the dilemma. To qualify as a farmer, the first requirement was land. Ghulam Hussain immediately donated a tiny plot of his own, large enough for his brother to keep a cow. With additional land and animals on it, Muhammad Hussain became technically a farmer.

Meanwhile, Salam had already set in motion the mechanism of the British university admission process, made additionally complicated by the remoteness of Britain, the end of a war and the volatility of the situation in India. Through the Secretary of Education in the London office of the High Commissioner for India (the equivalent of the Indian Embassy), Salam's name was put forward for admission to a Cambridge college, with recommendations from the stalwart Chowla and from Dr Abdul Hamid, the senior lecturer in mathematics at Lahore and the curator of Punjab University's astronomical observatory. At Cambridge, St. John's College had been expecting a graduate student from India who was going on to do research in English Literature, but who in August 1946 suddenly opted out. When the Indian High Commission offered an alternative research student, St. John's said it would now prefer an undergraduate. With his master's degree from Lahore, Salam could have applied to go directly into research at Cambridge, as Har Gobind Khorana had done in Liverpool, but with the family's sights still set on him entering the Civil Service, Mian Afzal Hussain, the Vice-Chancellor of the University of the Punjab, had advised Muhammad Hussain that his son should revisit undergraduate mathematics, this time Cambridge-style. There was still time – the ICS age limit was 25. In 1946, the idea of going on to do scientific research (as the future Nobel prizewinner Har Gobind Khorana had done from Lahore) had not yet occurred to Salam, to his later regret[17]. But all this lay in the future. When he received the Cambridge telegram in Multan on 3 September, Salam knew that he had to move fast, but was unsure exactly which direction to go.

With his university place now secure, he had to confirm that the Punjab scholarship money was available. On the same evening, Salam took the overnight train from Multan to Lahore to visit the offices of the Punjab Education Department, only to learn on arrival that the department had decamped to Simla, one of the lofty refuges built in the nineteenth century for a British administration crushed by the summer heat of India. Simla is 250 kilometres from Lahore, and the train has to grind up to an altitude of 2000 metres. When an anxious Salam finally arrived at the offices of the Education Department at 2 pm on 4 September, he was overjoyed to meet a messenger bearing a letter for him. Salam, along with four other Punjabi students, were being given money to go to Britain – 365 pounds per year for three years[18], more than his father earned. The offer was conditional on having a university place, but Salam had the telegram from St. John's in his pocket. He was the only one of the recipients of the post-war Punjab scholarships to obtain an immediate offer of a place at a British university. The other four 'successful' candidates had their scholarships deferred until they had a place, but meanwhile their money was swallowed up by other funds and the scholarships lapsed. As Salam said later 'The entire purpose of that fund and those scholarships seemed to be to get me to Cambridge'[19]. Later, Salam often mused on this eerie succession of strokes of fortune – the establishment of the post-war scholarships, his father's qualification through an accidental acquisition of land, and his admission to a college that would go on to play a major role in his life.

In Simla, Abdus Salam now had documents assuring him of a university place, and a scholarship. He was supposed to begin his studies in Cambridge in October. But first he had to get there. The boat trip to Britain would take several weeks. In Lahore he had already been told that it was 'impossible' to get a confirmed berth before the end of the year. With no time to lose, he took the train from Simla to Delhi, another 250 kilometres. By the time the harassed Salam arrived at the offices of the shipping company, it was Saturday afternoon, and they had closed for the weekend. After an argument, a reluctant clerk gave Salam the form to make a provisional booking. Salam then took the train back to Multan, where his father was waiting patiently at the station with a lantern. Because of Muslim–Hindu rioting, a curfew had been imposed in the town, but Muhammad Hussain had been given special permission to wait for his son. The entire day of 6 September was spent packing clothes and books, mostly the latter. A local notable, Malik Umar Ali, whose sons had been tutored by Salam, had given money for the coming journey.

Abdus Salam, aged twenty, said goodbye to his family, and departed with his heavy trunk for the port of Bombay, 1000 kilometres away, another two nights on the train. There he would wait for a place on a boat. Salam's father stayed at home to pray, convinced that this would achieve more than accompanying his son. On arrival in Bombay, the city was under curfew and Salam, exhausted by the train journey, bolted into a run-down hotel near the station. No sooner had he fallen asleep when he was woken by a pounding on the door. It was the British Military Police, looking for deserters from the Indian navy. Earlier that year, Bombay had been a flashpoint when the disgruntled navy had gone on strike. After producing his Cambridge University letters, Salam was able to go back to sleep.

The next morning he went to Bombay Docks with his provisional booking. A ship was sailing that afternoon, but getting on board would be a challenge. After a Labour government swept into power in 1945, British plans to move out of India had suddenly accelerated, precipitating a mass exodus of soldiers, administrators, and their families. The Bombay dockside was in tumult, with columns of British troops laden with equipment, and families clamouring for their baggage to be moved. Departing families were saying farewell to faithful servants they had known all their lives and who now had no future. Even with a heavy trunk, a solitary young man could weave through regiments of soldiers and entire families surrounded by their belongings. Emerging from the mass of transient humanity, Salam was assigned a berth on the 20 000-ton Franconia, bound for Liverpool. Built in 1922 for Cunard's North Atlantic route, the ship had been converted into a troopship in the Second World War: in February 1945, Churchill had used it as his floating headquarters at the historic meeting with Stalin and Roosevelt at Yalta, on the Crimean Black Sea coast[20]. As well as Abdus Salam, on 11 September 1946 the Franconia carried some 600 British families, leaving India with mixed feelings, and 600 liberated Italian prisoners-of-war, happy to be homebound.

Salam watched the shoreline recede. It would be several years before he would see the subcontinent again, by which time it would have been sundered into two countries. He shared a cramped berth with an Indian called Menai who was seasick for the entire 21-day journey. The ship's British food was disagreeable: bread, corned beef, and a soup that Salam's stomach was unused to, but at least he was in better shape than his cabin companion. After crossing the Arabian Sea, he bought his first wristwatch when the ship docked at Aden. He had learnt Arabic at school, but this was the first time he had heard it as a living language: street vendors speak rapidly and do not recite the Holy Qur'an. After transiting the Suez Canal, the Franconia called in at Naples, where the joyous Italian prisoners-of-war disembarked, and where Salam purchased a kilo of grapes to supplement the Franconia's monotonous menu. It was the first time he set foot in Italy, a country later to become his home. Several days later, Abdus Salam shivered as he watched the Liver Building and Liverpool docks loom through the early October mist. The shoreline looked cold and miserable. Having brought himself all the way from Multan, he was now ready for the last lap of his journey to Cambridge. He was unsure how to accomplish this in a cold and foreign land, but was comforted knowing that his family were constantly praying for him. After the hectic unpredictability of his journey so far, there was to be a pleasant surprise for him on the dockside. It looked as though the family prayers had worked.

REFERENCES

1. Singh, J., *Abdus Salam, a biography* (New Delhi, Penguin, 1992.)
2. Ghani, A., *Abdus Salam: a Nobel Laureate from a Muslim country*, (Karachi, published privately, printed Ma'aref, 1982)
3. Muhammad Abdur Rashid (Salam's brother) interview
4. see Ghani. A.
5. Khan, M. A., *Lifelong friendship with Abdus Salam* in Hamende, A. M. (ed.), *Tribute to Abdus Salam*, (Trieste, ICTP, 1999)
6. Originally a speech given in Urdu at Jhang College, 1972, subsequently published as an article in the Urdu monthly magazine of the Aligarh Muslim University, India, January 1986. Rendered into English on the Ahmadi Muslim Community website http://www.alislam.org/library/links/00000126.html
7. Salam, A., *The unification of fundamental forces, 1988 Dirac Lecture*, (Cambridge, CUP, 1990)
8. Salam, A., 1990
9. Salam, A., 1990
10. Nobel Prize lecture, 1979
11. Abdul Hamid (brother), English synopsis, dated September 1997, of Urdu biography,
12. Vauthier J., '*Abdus Salam, un physicien*', Beauchesne, Paris, 1990
13. Salam, A., *Homage to Chaudry Mohammad Zufrullah Khan*, in Lai, C. H. (ed.), *Ideals and realities* (Singapore, World Scientific, 1987)
14. Dalrymple, W., *The age of Kali* (London, Harper, 1998)
15. see also Ali, A., Isham, C., Kibble, T., Riazuddin (ed.) *Selected papers of Abdus Salam* (Singapore, World Scientific, 1994)
16. Ram Prakash Bambah, *Together in Lahore and Cambridge*, in Hamende, A. M. (ed.), *Tribute to Abdus Salam*, (Trieste, ICTP, 1999)
17. Salam, A., *A life of physics*, in Cerderia, H. A., Lundqvist, S. O., (ed.) *Frontiers of physics, high technology and mathematics* (Singapore, World Scientific, 1990)
18. see Ghani, A. (1982)
19. Wolpert, L., Richards, A. (ed.), *A passion for science*, (Oxford, OUP, 1988)
20. http://www.greatships.net/franconia2.html

5

From mathematics to physics

Unlike Oxford, a city big enough to engulf its university, Cambridge is subjugated by its ancient machinery of learning. Salam loved it, and flourished there. The university is an omnipresent network of some thirty colleges scattered about the town, each self-governing under a 'Master' and staffed by 'Fellows', who may also be university lecturers or professors. Each college selects its own undergraduates, and oversees their personal needs for accommodation, meals, and recreation, providing a social, as well as an academic focus. An important aspect of the college education is the close, almost parental, supervision given by tutors, covering general welfare as well as academic progress. However, programmes of study, lectures and examinations are organized by the University. Traditionally, academic staff worked at home or in their college rooms, but in the second half of the nineteenth century, the increasing sophistication and importance of science demanded central university laboratories and an astronomical observatory for teaching and research.

One of Cambridge's eternal strengths is its tradition in mathematics. From 1747, every Cambridge undergraduate had to undergo the fearsome mathematical 'tripos' examination, irrespective of their main subject of study. The name comes from the three-legged stool on which the official university examiner sat as he challenged students in fierce interrogation. In the late eighteenth century, the role of the 'tripos' examiner was downgraded, but the name stuck. In his autobiography '*Home is where the wind blows*'[1], Fred Hoyle wrote 'Mathematics had always been the jewel in the Cambridge academic crown'. The written mathematical exam, introduced in 1772, became a challenge in its own right, with the most able students competing for the accolade of 'wrangler', a term dating from when candidates had to argue with the tripos examiner. At the top of the pile sat the 'senior wrangler'.

The demanding mathematics tripos course in principle covers four years, including applied mathematics and theoretical science as well as pure mathematics. To cover a four-year period, examinations were divided into Part I and preliminary to Part II (now called Parts IA and IB), Part II and Part III. Most students get a degree after doing Part II in their third year, and only a few go on to do Part III in a fourth year. Bright students, or those already with a mathematics qualification, can skip some initial work, face Part II after two years and go on to Part III in their third year. Salam was to do otherwise.

Cambridge's tradition in mathematics became complemented by new knowledge in the sciences. In the nineteenth century, this advance was spearheaded by the Cavendish Physics Laboratory, founded in 1871 by the seventh Duke of Devonshire, the immensely rich Chancellor of the university. This new laboratory, which took the family name of the duke, was to become a scientific flagship of the twentieth century. Its first head was James Clerk Maxwell, who in 1864 had invented the new theory of electromagnetism, but had almost immediately retired from academic life at the age of 34 to his remote estate in Scotland. The idea of leading a modern laboratory at a prestigious university induced Maxwell to leave Scotland, and his arrival at the Cavendish Laboratory in 1871 set the stage for a century of remarkable achievement. In developments as far-reaching as those of Isaac Newton some two hundred years earlier, these would change forever our view of the world, opening up first the atom, then its nucleus, for closer inspection.

As the nineteenth century drew to a close, the world seemed to sense that the time had come for radical new ideas and innovations. The internal combustion engine provided a new means for mass transport. The telephone and wireless telegraphy revolutionized communications. Launched on the wave of impressionism, art was in the middle of an almost unprecedented period of originality and vitality. Freud was preaching a new understanding of the human personality. Physics too seemed to sense something just over the horizon. A contemporary viewpoint, exemplified by great nineteenth-century figures such as William Thomson (later Lord Kelvin), was that Nature organized itself like a Swiss watch, with tiny precision machines delicately interlocked in a vast cosmic scheme. The ultimate cogs in this mechanism, according to more audacious minds, were atoms (from the Greek meaning 'uncuttable') – indivisible pinpoints of the chemical that make up the material of our world. However, many contemporary scientists refused to believe in such atoms.

Ironically, evidence for atoms came when experiments showed that atoms were not indivisible and could instead fall apart. The audacious atomic picture was outmoded as soon as it appeared. J. J. Thomson (no relation to Kelvin), who had inherited the Cavendish professorship initially occupied by Maxwell, had supreme patience in handling delicate and temperamental apparatus. He watched carefully as a high voltage was applied across a glass tube containing gas at low pressure. Although the residual gas atoms in such a tube are electrically neutral, they are built up of electrically charged components that are stretched by the applied electric force. Such stressed atoms eventually break, releasing negatively charged 'cathode rays'. Thomson called these negatively charged atomic components 'electrons'. With atoms no longer the 'uncuttable' pinpoints of matter that their Greek nomenclature implied, attention shifted towards their inner structure.

If the electrically neutral atom contained negatively charged electrons, it also had to contain a compensating positive charge. The man who discovered it – the atomic nucleus – was Ernest Rutherford, J. J. Thomson's successor as Cavendish Professor. There are eerie similarities between the careers of Rutherford and Abdus Salam, both from far-flung parts of what was then the British Empire, and both accidentally winning prestigious scholarships to Cambridge. Rutherford's grandparents arrived in New Zealand from Scotland in 1843 to help establish a sawmill in the young colony. Born in 1871, the fourth of twelve children, Ernest Rutherford was soon seen to have ability, and was initially prepared, as Salam would be, for local Civil Service examinations before going on to college.

While Rutherford's parents had been growing up, on the other side of the world the 1851 Great Exhibition in London had been a success, both as a showcase for British achievement and knowhow, and as a financial venture. The profits had been ploughed into a new campus area in South Kensington, including great new museums and the nucleus of what would eventually become the Imperial College of Science and Technology. In 1891 the exhibition proceeds were also used to establish postgraduate research scholarships for students of outstanding ability. New Zealand was allocated about one such scholarship per year. For the 1895 award, based on research dissertations, there were two candidates. The local newspapers in Rutherford's town announced the eagerly awaited decision – 'The Science Scholarship is awarded to Mr J. MacLaurin of Auckland. Mr Rutherford of Christchurch was second.' In a two-horse race, Rutherford had come last. However, when Maclaurin read the small print, he realized that he no longer qualified. While several mail steamers had plied back and forth across the planet, MacLaurin had already obtained a New Zealand government job, and was no longer eligible for the award. Runner-up Rutherford stepped in and embarked for the UK in 1895. He was to be Cambridge University's first research student recruited from outside that university.

As Salam was to do half a century later, Rutherford arrived at Cambridge just as a powerful wave of scientific development was breaking on a new shore. Both rode them skilfully for the remainder of their careers. Rutherford's wave was the new insight into the atom, which he developed in a historic series of experiments, the greatest being his discovery that the cloud of negatively charged atomic electrons was electrically matched by a positive charge concentrated in a tiny nucleus deep at the heart of the atom. In Cambridge, Rutherford presided regally over a golden age of discovery, work that went on to earn a clutch of Nobel prizes in the 1920s and 1930s, and that established the Cavendish as the world's leading centre for research in nuclear physics.

The realization that atoms had a structured interior was a scientific revolution in its own right, but physicists soon found that the inner workings of these atoms undermined their smug nineteenth-century understanding. That complacent wisdom had been grounded in the grand system of mechanics that had existed since the time of Newton, and underpins the movement of celestial bodies, such as the Sun and its attendant planets, locked in the all-pervading grip of gravity. For tiny atoms built of a central electrically positive nucleus surrounded by a cloud of negative electrons, Maxwell's electromagnetism instead provides the motive force, the nucleus replacing the Sun, and the orbiting electrons the planets. But in this picture, electrons could not be locked in their atomic orbits for ever. An electron – a moving electric charge – should continually radiate, losing electromagnetic energy and eventually falling into the nucleus. For atoms that have existed since the dawn of the Universe, clearly this does not happen.

Atoms and their ilk are the denizens of an eerie quantum world that behaves in a way very different from anything we have directly experienced. The rethink was sparked by the arrival of the quantum picture at the outset of the twentieth century, when Max Planck and Albert Einstein showed that radiation energy is not a smooth stream, but is instead built up of separate packets, or 'quanta'. Light, which is one form of energy, arrives in the same way that rain falls as drops. We say that a certain place has on average 50 centimetres of rainfall each year, knowing full well that this is not a continuous current. But we have little idea what 50 cm of annual rainfall means in terms of raindrops falling per square centimetre. Civil engineers can design systems to control and channel rainwater without worrying that it falls as discrete drops.

However, physicists trying to understand the interior of the atom had to count their radiation quanta one by one, and over the years developed detailed accounting systems. This succession of quantum recipes is like the software continually being developed for today's personal computers. Each new version is heralded as a panacea, a cure-all for everything that went before, and is eagerly snapped up by technically literate users. But wider application of each new release soon reveals fresh deficiencies, and frustrated users eagerly await a software update, where the bugs have been fixed, only to reveal unsuspected new ones. A complete theory of the quantum world has yet to be developed. Instead, there has been a steady succession of more ambitious theories, each giving excellent results in its own limited domain until falling over the edge of its own limitations. It was in this development programme that Abdus Salam was to make his scientific mark.

The first such atomic quantum software came in 1913 when Niels Bohr showed how electrons do not orbit atoms anyhow, but are instead locked into definite paths, or orbits. However, when a Bohr electron is nudged hard enough, it can make a 'quantum jump' from one orbit to another, in the same way that climbing up or down a ladder requires each foot to be moved high or low enough to encounter the next rung. In the early 1920s, physicists learned how to account for these jumps, and the behaviour of simple atoms like hydrogen could be worked out mathematically. How one did so was initially a matter of taste: there was the 'wave mechanics' of Erwin Schrödinger and the 'matrix mechanics' of Werner Heisenberg. Both systems worked, but their interrelation was less clear. It was like the Gospels – one theme viewed from different perspectives. The man who saw the meaning of these quantum perspectives was Paul Dirac in Cambridge.

In 1664, when a new Professorial Chair was established by Henry Lucas, the member of parliament for the university, Cambridge appointed its first professor of mathematics – Isaac Barrow of Trinity College. Barrow had begun his academic career with Greek and theology, continued with medicine, and finally added astronomy and geometry. One of his early duties had been to assess the mathematical abilities of a young student called Isaac Newton. The unconventional Newton had hardly bothered with formal texts, and his undergraduate results suffered. In 1665, an outbreak of plague closed the university, and the students dispersed. By the time he returned to Cambridge in 1667, a silent Newton had totally rewritten contemporary mathematics and science. One of the few to realize was Barrow, and in 1670, Newton, aged 28, succeeded as Lucasian Professor.

In 1932 the chair was inherited in turn by Paul Adrien Maurice Dirac. Born in 1902, the son of a Swiss immigrant to Britain, Dirac became a legend for taciturnity as well as intellect. Fred Hoyle (of whom more later) said of Dirac 'More than any other person I have known, Dirac raised the meaning of words and syntax to a level of precision that was mathematical'[2]. During the Second World War, a Minister, anxious to recruit Dirac to the war effort, asked him to call in 'next time he was in London'. Dirac assured the Minister he would. As an afterthought, the Minister asked how often Dirac came to London, to which he replied 'About once a year'[3].

As a Cambridge research student in the mid-1920s, Dirac, worried by the apparently conflicting quantum pictures of Schrödinger and Heisenberg, sat down and carved out his 'Principles of Quantum Mechanics', which for the first time explained the new quantum mechanics and expressed it in a self-consistent mathematical form. Although it is about physics, it has no diagrams, refers to no explicit experiments, has no references, and contains no suggestions for further reading. After Cambridge University Press had dismissed the book as unpublishable, Oxford stepped in. Ever since, it has been the classic introduction to quantum mechanics for generations of students, and remains a continual source of inspiration for researchers. It is often compared to Isaac Newton's 'Principia', another book much more widely acknowledged than actually read. However, while Professor Newton was sidelined as a crank and disregarded by students, sometimes lecturing to the walls, Dirac was an inspiration, his book providing the script for his lectures. This was to be Salam's introduction to quantum theory. Two contemporaries of Salam, Richard Eden and John Polkinghorne, wrote 'Dirac's greatest influence on students at Cambridge…was through his course of lectures on quantum theory. For many years it was the first course in quantum theory that

Cambridge students could attend. However, not all the audience were novices, for frequently visitors of some standing would rightly judge it not to be missed. There was more to the lectures than the printed page can convey. One was carried along in the unfolding of an argument that seemed as majestic and inevitable as the development of a Bach fugue'[4].

The new quantum theory was one of the twin pillars of the new twentieth-century physics. The other was Albert Einstein's special theory of relativity, which pointed out the dramatic implications of the velocity of light emitted by a source at rest being the same as that emitted by a moving source. Dirac's greatest achievement was in 1927 when he wrote down an equation of breathtaking conciseness and symmetry that linked the quantum electron with such relativity. It predicted that the electron had to have a counterpart particle, called by Dirac the 'antielectron', but subsequently renamed the positron. The discovery of this antiparticle mirror world in 1932 was an impressive demonstration of the power of a single equation. In 1933 Erwin Schrödinger, Werner Heisenberg and Paul Dirac received the Nobel Prize for Physics[5].

Despite their proximity at Cambridge – Dirac at St. John's College and Rutherford at the Cavendish – and their Nobel prize reputations, the careers of Rutherford and Dirac hardly touched. The ultimate irony came in 1932, when the antielectron which Dirac's equation had predicted was discovered in California, not Cambridge. While Rutherford had reigned magisterially over his Cavendish laboratory empire, Dirac sat alone in his college study. His 'workshops' were long solitary walks. Dirac rarely accepted individual students. When one aspirant asked to be taken on, Dirac replied 'I am very sorry, but I don't think I need any help with my problems at the moment'.

One student who was taken on by Dirac was Fred Hoyle. Born in West Yorkshire in 1915 in a modest but accomplished family, Hoyle was something of a British counterpart to Richard Feynman. Both were gifted scientists and individualists with fierce temperaments; both came from modest backgrounds; both had supreme ability; and both had their research careers interrupted by the Second World War . Both brandished fierce accents (Feynman's from New York City, Hoyle's from Yorkshire) as personal trademarks. Both were gifted communicators; and both could be irreverent of authority.

Hoyle rose to become Plumian Professor of Astronomy at the University of Cambridge, a grand title. But he was also a prolific popularizer of science and writer of science fiction, and such diverse achievements were sometimes difficult for his colleagues to reconcile. His extrovert behaviour, straight talking, and success as a writer and broadcaster continually irritated some of his influential contemporaries, affecting his Cambridge career and possibly his Nobel Prize chances as well. Furious at a series of adverse decisions, Hoyle threatened several times to resign, and finally departed in 1972 after being again overlooked for a new appointment, walking out of the university in a final thunderclap of frustration[6].

As a novice research student, Hoyle's first research supervisor had been Rudolf Peierls from Germany, who had worked with quantum mechanics pioneers before fleeing Nazi persecution in 1933. Settling in Cambridge, Peierls and his wife Genia had taken the bachelor Dirac as a boarder. In 1938 Peierls moved to the University of Birmingham, leaving Hoyle on his own. By this time, Hoyle's research was going well, and supervision was not needed. But for administrative reasons, a student had to have a formal overseer, so Hoyle quickly had to find one. The solution was Dirac, and the duo were happy with the ironic circular logic of a student who did not want a supervisor being assigned to a supervisor who did not want a student. Then came the war. Hoyle's research career still lay in the future when he returned to St. John's College, Cambridge, in 1945, one year before Abdus Salam arrived as an undergraduate and was assigned to Hoyle for mathematics tuition.

Few Indians arrived at British immigration queues in those days. As British subjects, at that time they were in theory free to travel anywhere in the British Empire, as long as they could fend for themselves. Abdus Salam's arrival had been smoothed by a fortunate meeting at the dock in Liverpool. Sir Muhammad Zafrullah Khan, then 53, was the most prominent Ahmadi in India, at the time one of the Judges of the nation's Supreme Court. Soon he would be invited by Muhammad Ali Jinna, the *Qaid-i-Azam* (Great Leader) of Pakistan, to become Foreign Minister of the new nation. Later he would also play a key role in Salam's international career, but in October 1946 he was in Britain, *en route* from the United States, and had come to the blustery Liverpool dockside to welcome a nephew. Salam had once seen Zafrullah Khan as a distant figure in 1933, but this was their first actual contact. Salam stood with his huge trunk of mathematics books, well prepared for university study, but was totally unprepared for the British climate. He shivered in the early autumn chill. Seeing the student's predicament,

Zafrullah Khan gave him his heavy overcoat and helped him with his luggage.

Wrapped in the coat, Salam took the train with Zafrullah Khan to London. During the journey, the lawyer pointed out landmarks, but Salam was more impressed by the greenness of the countryside. He stayed overnight with Zafrullah Khan at the Ahmadi London Mosque in Southfields, before making his way to Cambridge. From the station, he took a taxi to St. John's College. Getting out of the taxi with his 40-kilogram trunk, he asked the college porter for help, and was shown a wheelbarrow. Unlike Bombay docks, there was nobody to carry loads. Salam was so overcome at arriving at his destination that he forgot to ask where his rooms were. He learned that he was in New Court, whose intricate architecture could have reminded him of the Mughal splendour of Lahore. He was allotted three buckets of coal each week. But coal was not his first requirement, and the depths of the winter were yet to come.

Despite having to haul his own luggage, this was a soft landing for a confused young foreign student: Britain was no longer the country that Salam had heard and read about. At the beginning of the century, Britain had been master of the world, with London as a global capital. But in 1946, mortgaged to two world wars and with its Empire collapsing, post-war Britain was only a shadow of what it had been. Uncomfortable in their 'demob' suits (every demobilized soldier was given clothing for civilian life), fathers with children they hardly knew tried to pick up the threads of their former lives. 'Prefab' bungalows, hastily assembled on weed-covered sites, provided substitute homes for those whose houses had been bombed. In this sequel to the war, the Socialist government had introduced radical new measures that affected everyone's lives: great industries that employed a major fraction of the nation's workforce had been nationalized, and a new National Health Service promised free medical treatment for all.

The most essential personal possession was a 'ration book' that had to be presented for each purchase of meat and groceries. Each person had coupons for enough meat and cheese for about three days a week, together with limited amounts of tea, bacon, ham, butter, sugar, margarine and tea. Food was fried in generic 'cooking fat', also rationed, as were eggs, but passable scrambled egg could be concocted from a dehydrated powder, made in the USA. Coffee, virtually unobtainable, was replaced by synthetic products, such as a dark essence called 'Camp', which was diluted to taste with hot water. Potatoes, freely available during the war, were now also rationed. But nobody starved: rationing simply meant there was little or no choice, and the resultant monotony clouded the gastronomic memory of several generations of Britons. Apart from the bomb sites, the streets were unchanged from 1939: with clothing a bare necessity, fashion had not altered, and the same cars and public transport vehicles, seven years older, were still on the streets.

With the British still wearing patched pre-war clothes, Abdus Salam's Lahore tailoring was not evident. He spoke English well, but with an accent. His skin was dark. He was different. In India, he had seen and known Britons, but they were one in a thousand: officials, teachers or missionaries. Now they surrounded him, their skins no longer reddened by semi-tropical sun, eating their sandwiches on railway platforms and doing even the most menial jobs. It was difficult for him to tell the difference between a professor and a porter, especially when they wore the same college tie. It is ironic that the inhabitants of countries where the skies are grey accentuate this drabness by their clothing, while those in more tropical countries choose bright colours. The drabness of post-war Britain was reinforced by clothing rationing. To someone who washed five times a day before praying, British people who washed once a week must have appeared dirty and smelt unpleasant. After using the toilet, they 'cleaned' themselves with crinkly paper, or even newspaper. In the subcontinent, the custom is to use water.

To a young student from another country, even with a good command of English, the most mundane things can create misunderstandings. Greetings are a social minefield. Britons do not bow. When is it appropriate to shake hands? When a shop assistant or a ticket clerk addresses a client as 'love', or 'dear', this does not imply a proposal of marriage. Smiles to strangers as a polite reward for services rendered have to be precisely regulated to an appropriately feeble wattage. Close body contact on crowded public transport is unfortunate but has no sexual overtones. In their cities, Britons did not speak unless spoken to, and then only for a clear reason. The British had, moreover, just endured a major war by stoically accepting whatever came their way. 'Musn't grumble,' was the traditional reply to 'How are you?', even if some tragedy had befallen. For Salam, many of the everyday habits and customs of daily British life appeared bizarre, much as beggars and cows in the streets do to tourists in India, to be avoided but never forgotten.

Someone from the Indian subcontinent working at Cambridge University was shielded, but not isolated, from any wind of British colour prejudice. At St. John's, he was called 'Mr Salam' or even 'Sir'. But prejudice at a personal level within the university lurked below the surface. In being assigned to Fred Hoyle's able tutorship, Salam was fortunate. Hoyle enjoyed working with intellects from the subcontinent, later collaborating with Jayant Narlikar and Chandra Wickramasinghe. At Cambridge in 1946, Indian students were a small but identifiable minority, large enough to provide each other with mutual support. Spread over the university, each college averaged about one entrant per year from the subcontinent. But then Salam was Muslim, and knew what was happening back in his home country. How would this affect his contact with Hindu students?

More important was the contact with British students. Salam already had two university degrees. His fellow mathematics students, most of whom were studying for their first degree, spoke English without an accent. Many of them had just been on active service in a war and had experienced matters of which Salam knew little. Most of them also relished sport; rugby, rowing, athletics... foreign to the totally sedentary Salam. Their booming voices and boisterous self-confidence made Salam assume they were all intellectual giants, twentieth-century Newtons in the making. He was impressed by their respect for their teachers: the total 'pin-drop' silence during lectures reminded him of prayers at a mosque. He was impressed by his contemporaries' meticulous note taking during lectures, using rulers to draw straight lines. He also noted the Cambridge emphasis on respect and self-reliance.

After the uncertainty of a long voyage, Salam's found St. John's a delight. Founded in 1511, the College has a privileged position, backing onto the river, but in the centre of the city, just north of King's and Trinity Colleges. Among its famous graduates had been the poet William Wordsworth. Salam's rooms in New Court (built in the nineteenth century) were larger and better furnished than the family home in Multan. They were cleaned by college servants who would also make his bed and shine his shoes each morning, and he would eat his meals in the college hall. After his introduction to college food at Cambridge, for the rest of his life Abdus Salam enjoyed eating in student cafeterias and refectories. The college rose gardens were a delight after the heat and dust of the Punjab. His scholarship was worth about £365 a year, generous when anyone earning £10 a week in Britain was presumed to be well-off and was therefore heavily taxed, and was sufficient to subsidize his family in Multan. Salam was impressed by the discipline, assumed rather than imposed, with students encouraged to be in their rooms by 10 pm, but with post-midnight returns not tolerated.

With four lecturers, St. John's was well suited to mathematics students. Salam also discovered the delights of its well-stocked library. 'Specialization in one area is the sterilization of one's intelligence,' he said later[7]. As well as his mathematics studies, he read all he could find on religion, and learned much about history, studying for up to sixteen hours a day. Soon he was joined at Cambridge by Ram Prakash Bambah, who he had known at Government College, Lahore. When they were tired of study, they would walk together through the town, go to the cinema, or talk Punjabi in their rooms until late. (St. John's had a tradition of taking talented students from the subcontinent. A decade after Salam came Manmohan Singh, later to become India's fourteenth prime minister. Salam's brother Abdul Majid arrived in 1952 to study natural sciences.)

Because Salam already had a mathematics degree, he was launched directly into the preliminary course for Part II, bypassing Part I. He also arrived in the UK just in time to experience the exceptionally bitter winter of early 1947. Later, Fred Hoyle wrote[8]

'Unless you actually experienced those post-war years, it is surely impossible to visualise how bad it really was. Churchill had promised us an ascent to the "broad sunlit uplands". What it actually brought was a descent into that appalling winter.

I had rooms in New Court which by common consent was the worst place in College to be. Designed with mid-nineteenth century spaciousness, rooms had been planned with large fireplaces and wide chimneys that gobbled coal. [They were] designed to be lit in the morning by a college servant and "made up" similarly throughout the day. But definitely not in 1947. If you wanted a fire that winter you lit it yourself, and if you wanted it "made up" you made it yourself. Except you didn't because you were out of coal. My ration for a whole week's supervision of mathematics students in I8 New Court was one bag per week.

The one luxury we enjoyed was non-material. John's still maintained four College Lecturers in Mathematics. There was Peter White and Frank Smithies on the "pure side" and, Leslie Howarth and myself on the "applied side". Howarth had the rooms next to mine, and technologically more advanced. The fireplace had been blocked and Howarth luxuriated in a gas fire, of which the operative area of the front measured about 4 × 8 inches.

Off-peak it would glow a fairly bright red but those who arrived at noon for a supervision with Leslie would be greeted by a small rectangle of pale pink.

Like everything else, clothes were severely rationed. Meaning that it was a case of wearing whatever you could lay your hands on, no matter how outlandish the garments. Anyway my students soon learned to pick on whatever they could seize hold of, as we sought earnestly to solve ingenious problems of spheres rolling on spheres. It was into this icy atmosphere that Abdus Salam found himself plunged on his arrival in Cambridge. Much of his later success can be attributed to the fact that he survived it.

Salam had already done a mathematics degree in India, as it would have been then, Pakistan as it was to become. Warmth into cold he must have expected. But food into no-food he did not. His food ration book would have been taken immediately on his arrival in College. Through that first winter, he always averred in later life, he lived on apples, which was all that were in the markets to be bought without coupons. Except perhaps potatoes with which I suppose he was not equipped to cope. No joke this – even as late as 1951–52 the weekly British cheese ration was a mere one ounce. Why people stood for it tells a not particularly flattering story about the British temperament.

As the senior of the four College Lecturers in Mathematics, it fell to Peter White to decide how to group students. A grouping into an occasional one, but mostly in twos and sevens, lasted generally for a year. Occasionally there would be a permutation that produced a minor shift but not often. Each student got two hours supervision each week, one hour of pure and one of applied. And there was an alternation term-by-term between White and Smithies on the pure side and Howarth and myself on the applied side. It was a system that put as little strain as possible on the individual College Lecturer.

Anyway, Abdus Salam was one of the rare ones who had to be 'taken' alone, there being no obvious partner or partners with whom he could be grouped. Howarth had him in the first term of his first year. Howarth told me over coffee one night after dinner that he had a "man from India who was very good", which was the first I ever heard of Abdus Salam. What I also heard about him from Howarth was that he had the embarrassing habit of greeting [others] in the John's Courts with a fully pledged Muslim salute, practically going down on the cobblestones with his knees. It must have taken for Leslie, or for Peter White I suppose, to inform him that such reverence was not considered necessary in Cambridge. At any rate the full Muslim greeting had been reduced to a wave of the arm and a shout by the time it came to my turn to have him for supervision on a one-to-one basis.

It was then when the real cold struck, with matters reduced to plain survival. I would be anticipating the end of the hour, when it would be possible to rush to another room, where an austerity fire would be burning, and Salam would no doubt be anticipating his next apple.'

Underlining Hoyle's description of 1947's winter of discontent, the British Meteorological Office says[9]:

'From 22 January to 17 March 1947, snow fell every day somewhere in the UK, with the weather so cold that the snow accumulated. The temperature seldom rose more than a degree or two [Fahrenheit] above freezing. Across Britain, drifts more than five metres deep blocked roads and railways. People were cut off for days. The armed services dropped supplies by helicopter to isolated farmsteads and villages, and helped to clear roads and railways.

In mid-January 1947, no-one expected the winter to go down in the annals as the snowiest since 1814 and among the coldest on record. The winter began in earnest on the 23 January, when snow fell heavily over the south and south-west of England. The blizzard in south-west England was the worst since 1891; many villages were isolated. The cold, snowy weather continued. February 1947 was the coldest February on record in many places and, for its combination of low temperatures with heavy snow, bore comparison with January 1814. The mean maximum temperature for the month was 0.5 °C (6.9 °C below average) and the mean minimum was –2.7 °C (4.6 °C below average).'

Another unusual feature of February 1947 was the lack of sunshine. At Kew (London), there was no sun on 22 of the month's 28 days. When skies did clear, night-time temperatures plunged. A minimum of –21 °C was recorded. In some parts of the British Isles, snow fell on as many as 26 days in February. Much of the snow was powdery and was soon whipped into deep drifts by strong winds. March was even worse. In the first half of the month, there were more gales and heavy snowstorms, with drifts five metres deep in the Pennines. This was compounded by fuel shortages and transport problems. Experiencing his first British winter, Salam must have thought it normal, and accepted the cold with fortitude and all the clothes he could find. Zafrullah Khan's overcoat must have been a godsend.

As winter finally relented, Salam, trying to concentrate on his studies, was distracted by alarming news of what was happening in the Punjab. As the strain of imminent partition mounted, Hindus and Muslims had turned on each other again, this time more seriously as the Punjab was torn apart. In Multan, thousands were killed. Worried about his family and guilty at his isolation, his conscience nagged him to return, to serve his new country as best he could, just as his fellow undergraduates had done for Britain in their war. Salam applied for a St. John's travel grant, but without success. Salam's father sternly instructed him to stay where he was and complete his studies. Bambah and his other Indian friends also told him to stay. To get on its feet, the new nation of Pakistan would need skilled administrators. The traditional family goal of entering the Civil Service was back on track, now with a new country in mind.

Hoyle continued: 'The winter finally gone but not forgotten, it was mid-June before I saw Salam again. I asked him how he had done in 'Prelims'. He said awful, with a lot of absurd mistakes and then disappeared with a big laugh. In the case of Prelims the class list plus actual marks was sent round to supervisors. Salam had a first and was, I believe, third on the list.' For his achievement in the first year examination, Salam averaged over 80% in the four papers, and was awarded a College scholarship, worth £60 a year. Mushtaq Ahmed, Imam of the London Ahmadi Mosque, signed the formal papers on behalf of Salam's parents. Salam was happy, but not really surprised after having done several years of mathematics at Lahore under Chowla. What did puzzle him was that he had done so much better than all the potential Newtons around him. Asking his college tutor, J. M. Wordie, a geologist later to become master of St. John's, he was told that this was the whole point of the examination, to take the sting out of those who took themselves so devastatingly seriously[10].

In his second undergraduate year, Salam left the college for lodgings in town. This meant he had to stand more on his own feet, but this was easier after a year's experience of Britain. Salam still rejoiced in his work and in the college library, but in the long vacations, the college grew silent, especially during the cold, dark month of December. Salam knew nothing of Christmas until befriended by a British student, Christie, whose father was a railway engine driver in Shoeburyness, on the Essex coast. There, in December 1947, Salam experienced the warmth of Yuletide. For his second-year studies, he should have followed the lectures for Part II of the mathematics course, but emboldened by his first-year performance and by his solid mathematics grounding from Lahore, chose instead to follow the more advanced lectures for Part III. For this, he learned general relativity and gravity theory from Hermann Bondi, Hoyle's research collaborator. In Paul

Dirac's classic lecture course on quantum mechanics, the unfolding majesty of its logic was a turning point in his education.

But Salam did qualify in the summer of 1948 for a trip to Germany, organized by the US Control Commission and the *Bayrischer Jugendring* for students from Cambridge and several other European universities. The visiting students were housed in huge tents in Munich's city park. Despite the war having been over for three years, Salam was shocked by the extent of the destruction, with the inhabitants appearing to live in 'pigeon-holes'. He learned that a German was looking for him. The Bavarian had learnt Punjabi from mostly illiterate Indian soldiers and was trying to compile a prototype German–Punjabi dictionary. He had procured Waris Shah's Heer Ranja and some other texts, and asked Salam for help with some difficult passages. Salam did what little he could, and the memory of the German's diligence in the midst of such misery remained with him[11]. Later, Salam wrote about the trip in the Ravi, the magazine of Government College, Lahore[12]. He described the German food supply as 'desperate', with apparently only 50 pounds of dark bread per month to keep the people alive. Hospitality for foreign guests appeared frugal, but Salam learned that his daily serving of cheese was the equivalent of a week's ration for a German. He met refugee workers earning 150 marks per month clearing brickwork, having to pay 95 marks for basic food and accommodation. The Germans, he said, were pessimistic about the future, but optimistic about the present, never missing the slightest opportunity to enjoy themselves by singing.

In 1948, despite having attended Part III lectures, Salam was a wrangler, a first-class in the Mathematics Part II examination, effectively achieving in two years what most students do in three, and better. He had done what had been expected of him academically. He wrote again to his father, asking whether he should now return home to do what he could to help, despite the fact that his scholarship still had one year to go. Muhammad Hussain sought the advice of Ahmadi leader Hazrat Bashiruddin Mahmood Hussain, who said it would be an act of 'cowardice' if Salam abandoned a year of scholarship to return home[13].

So Salam began his mathematics Tripos year. Hoyle continued:

'I seemed to see much more of him. It was a clash of two cultures. Back home he had been educated in what might be called the Ramanujan school, according to which knowing what is true takes first priority, with knowing how to prove it a definite second, while I had absorbed the Cambridge system in which knowing what is true is not seen as of much relevance, only knowing how to prove it. Between us we managed to solve most Tripos problems.

I found it much less of a strain to tackle hard problems with a student like Salam than it was to be asked easier things by those chaps who just sat there and stared out into space. With the latter you had to roll two stones uphill simultaneously. One stone was the problem itself, the other was to get the chap to understand. With Salam you only had one stone and he would do a fair amount of the pushing.

After Salam did what was expected, first-class in Part II of the Mathematics Tripos, I ran into him again, this time in Third Court. He gave me his big hail. He had a problem he said, a policy problem: the people back home, Pakistan now, had granted him a scholarship for a third year. He had a thought that he might take Physics Part II, rather than Maths Part III. But not having 'done' any experimental physics to this point, he could hardly expect to achieve better than an Upper Second. Whereas if he went for Maths Part III he felt reasonably confident of a first, which would be much better received by the authorities back home. What did I think he should do? After some discussion I eventually said he should do what he judged would be best for Pakistan in the long run, rather than being too much concerned by short-term judgements, which I rather thought meant he should do Physics Part II. In after years he always said this was the most critical conversation of his life.'

Salam was walking the tightrope of his conscience. The Pakistan Civil Service was still his official intention, declared to his father. But his introduction to the quantum world in Paul Dirac's lectures in the Mathematics Part III course had impressed him deeply. Paul Dirac was a Fellow of St. John's College, where Salam would see him from afar several times a week, eating at High Table. Salam felt himself pulled in a new direction, and a new personal goal distilled from the family Civil Service objective. Salam was now drawn towards research in theoretical physics. He wanted to emulate Dirac, to carve mathematical monuments to Allah's work.

Mathematics he knew well, now having followed two full undergraduate courses. But he knew little of physics. Hoyle told him 'if you want to become a physicist, even a theoretical one, you must do the experimental course at the Cavendish Laboratory. Otherwise you will never be able to look a theoretical physicist in the eye'.[14] Salam went to his college tutor and explained his new plan. Wordie 'rubbed his hands in glee', relating how both G. P. Thompson (the son of J. J. Thompson) and Nevill Mott had tried to get a first-class result in Physics Part II in just one year after Mathematics Part II. Both had failed, but G. P. Thompson had shared the Nobel Prize for Physics in 1937, and Nevill Mott would emulate this feat exactly 40 years later. Doing Cambridge physics finals after just one year was harder than getting a Nobel Prize. As an 'experiment', Wordie put Salam's name down for Physics Part II, to see if he could do better than Thompson and Mott. The Nobel Prize could take care of itself.

Experiment was indeed the theme that year. Salam said later 'It was hard doing experimental work at the Cavendish, the hardest year of my student days.'[15] 'In the Cavendish, there was ancient equipment, nothing but. Rutherford's own equipment. And one was supposed to make it work. One had to blow glass and carry it three flights of steps. It was torture. They wanted it to be.'[16] In his first laboratory undergraduate project, Salam had to measure the difference in wavelength between two lines in the spectrum of sodium. He knew what the answer should be, took three readings, plotted them on a graph, drew a straight line through them and after three days dismantled his apparatus. His work had to be presented to a fearsome teacher – Denys Wilkinson, later to be Professor of Experimental Physics at Oxford from 1959 to 1976. Salam proudly displayed his straight line. Wilkinson looked at it disapprovingly and asked Salam what his background was. When Salam revealed he was a mathematician, Wilkinson replied 'I thought so. You realize that instead of taking three readings, you should have taken a thousand. This is just not worth grading.' In another painful experiment, Salam was supposed to achieve laminar flow in a set of glass tubes he had blown himself, but the tubes became clogged up and there was no flow at all. However, laboratory work at Cambridge did have some attractions. Salam appreciated the opportunity to work with the historic equipment at the Cavendish that Maxwell had used almost a century before to compare electric and magnetic measurements[17]. However, when it came to the laboratory work for his final examination, all did not go well. Salam had to be with his experiment for eight hours, and took chicken sandwiches for lunch. At the end of the long day, writing up his results, he suddenly realized he had used an inappropriate procedure. It was too late to do anything. Salam glumly turned in his results and rushed back to his room, where he immediately wrote to his father and asked him to start praying.[18]

Salam had tried to avoid Wilkinson after their initial laboratory encounter, but bumped into him again when the results of the Part

II Physics examination were posted. As Salam anxiously peered at the noticeboard, Wilkinson arrived. Looking over Salam's shoulder, he asked 'What [result] have you got?' As modestly as he could, Salam informed that he had a first-class degree. Salam relates '[Wilkinson] turned full circle on his heel, three hundred and sixty degrees, and said "Shows you how wrong you can be about people".' Salam was probably just as surprised as Wilkinson. J. M. Wordie's suggestion had worked. So had Muhammad Hussain's prayers, underlining the family's fervent belief in the power of pious devotion.

In 1949, after an absence of three years, it was now time for a maturer Salam to return home. But he could not 'return' to Pakistan, as he had never been there: it had only come into existence in 1947. His father was not rich and was reaching retirement age. Pakistan had no social security system: the family was supposed to provide it. Salam was still undecided whether to follow his father's wishes and try to enter the civil service, or go for his new ambition of research. Bambah tried to dissuade him from the former (Bambah eventually returned to India, becoming Professor of Mathematics at Punjab University, Chandigarh). 'Pakistan has many people who would make good administrators, maybe better than you, but it's hardly likely that there'd be anybody who could make [your] contribution to science.'[19] Several days later, Salam told Bambah that he had packed his belongings. He would leave the heavy trunks with his colleagues and return to Pakistan. If he found some way of supporting himself and his family, and that would also allow him to return to St. John's, he would come back and claim the trunks. Otherwise his colleagues were to ship them to Pakistan.

In Pakistan, Salam had another important assignment: he married Amtul Hafeez Begum, the second daughter of Ghulam Hussain, his father's elder brother, and a fomer student at the Ahmadi school in Qadian. In the Punjab, marriage between first cousins was, and still is, widespread. Islam entitles women to inherit property, so marrying a cousin ensures this remains within the family. Abdus Salam and Amtul Hafeez had known each other all their lives: their marriage in Jhang on 19 August 1949 was a joyous occasion, cementing the ties in a family that was already very close. Their 'honeymoon' was with Salam's family in Multan.

Armed with his impressive Cambridge results, the newlywed Salam was immediately employable in Pakistan. Mian Afzal Hussain, Chairman of Pakistan's Public Service Commission, informed him that the door was now open to the civil service, but Salam was by now adamant that he wanted to return to Cambridge, where his trunks still waited for him, and embark on a research career. For this, he asked Afzal Hussain for an extension of his government scholarship, to supplement a modest studentship already awarded by St. John's on the strength of his exam results. The science office in Karachi agreed, but the award, for two years, was not enough to support a wife, and after six weeks Salam returned to Cambridge, alone, leaving his new bride in the family home. He would not see her again for three years. Before his trip back, the prescient Salam also sounded Government College, Lahore, about future career prospects.

Having succeeded where G. P. Thomson and Nevill Mott had failed, obtaining a first-class physics degree in one year, Salam was fast-tracked by Cambridge for research work in the Cavendish Laboratory, then in the twilight of its contributions to front-line subnuclear physics, but still with an illustrious tradition. Salam now believed he had arrived at his final career path, but via a long and roundabout route. If he had not been pressured to enter the Civil Service, he now saw that he could have gone on to do this research at Cambridge more or less directly after his mathematics studies in India, as Chandrasekhar had done in 1930 and Khorana in 1946. Much of Salam's Cambridge undergraduate mathematics had duplicated what he had already covered. 'It would have been better if I had gone on to research right away – provided I had wanted to,' he admitted later[20].

The turning point had come with Dirac's lectures in his second undergraduate year. In his obituary for Salam in 1996, Hoyle said 'For Salam, the greatest scientist of the twentieth century was undoubtedly Dirac. Of course, you could say this was one John's man supporting another. But when I asked [Salam] if this included Einstein, he was clear in his answer, which went something like this: 'Einstein had his mathematics all done for him. Dirac invented his. Not only that, but it was Dirac who first made it clear that the route towards real understanding in theoretical physics lies through abstract mathematics, not through engineering mathematics.' For those of us who do not aspire to more than engineering mathematics this may seem deflating. But I think it was entirely correct.'[21] Later, Hoyle became a frequent visitor to Salam's Institute of Theoretical Physics in Trieste. But that lay far in the future. In the summer of 1949, Salam thought he knew at last in which direction he had to go.

REFERENCES

1. Hoyle, F., *Home is where the wind blows,* (Mill Valley, CA, University Science Books, 1994)
2. Hoyle, 1994
3. Hoyle, 1994
4. Eden, R.J. and Polkinghorne, J. C. *Dirac in Cambridge*, in *Aspects of quantum theory*, ed. Salam, A. and Wigner, E., (Cambridge, CUP, 1972)
5. Heisenberg received the Nobel Physics Prize for 1932, and Schrödinger and Dirac shared the 1933 award, but both awards were made in 1933.
6. Mitton S., *Fred Hoyle, a life in science* (London, Aurum Press, 2005)
7. Vauthier, J. *Abdus Salam, un physicien* (Paris, Beauchesne, 1990)
8. Sir Fred Hoyle, obituary to Abdus Salam, *The Eagle*, St. John's College, Cambridge, 1997, 80–5, published with permission
9. http://www.metoffice.com/education/secondary/students/winter.html
10. Salam, A., *A life of physics*, in Cerderia, H. A., Lundqvist, S. O., (ed.) *Frontiers of physics, high technology and mathematics* (Singapore, World Scientific, 1990)
11. From an article in the Urdu monthly magazine of the Aligarh Muslim University, India, January 1986. Rendered into English on the Ahmadi Muslim Community website http://www.alislam.org/library/links/00000126.html
12. Salam, A., A Visit to Germany, The Ravi, XLV, November 1951, Government College, Lahore.
13. Abdul Hamid (brother), English synopsis, dated September 1997, of Urdu biography
14. Salam, A. *Science sublime*, in *A passion for science*, Wolpert, L., and Richards, A., (ed.) (Oxford, OUP, 1988)
15. Salam, A. *Science sublime*, in *A passion for science*, Wolpert, L., and Richards, A., (ed.) (1988)
16. Salam, A., *A life of physics*, in Cerderia, H. A., Lundqvist, S. O., (ed.), 1990.
17. A Salam, *Unification of Fundamental Forces*, 1988 Dirac Memorial Lecure, CUP, Cambridge, 1990.
18. Abdul Hamid (brother), English synopsis, dated September 1997, of Urdu biography
19. Bambah, R. P., *Together in Lahore and Cambridge*, in Hamende, A. M. (ed.), *Tribute to Abdus Salam*, (Trieste, ICTP, 1999)
20. Salam, A., *A life of physics*, in Cerderia, H. A., Lundqvist, S. O., (ed.), 1990.
21. Sir Fred Hoyle, obituary to Abdus Salam, *The Eagle*, St. John's College, Cambridge, 1997, 80–5, published with permission

6

The men who knew infinities

A single electron, described by Dirac's designer equation, is not a museum piece to be admired inside a showcase. Electrons are constituents of our world, cogwheels in an atomic mechanism that are invisible, but make our world recognisable – sunlight, colour and warmth. All that is familiar and comfortable is due to electrons. The motive power for these electronic cogwheels is provided by the electric and magnetic effects that Maxwell had wrought together in his unity of electromagnetism.

When a cog in a mechanical motor turns, it emits a smooth hum when the motor is working well, increasing in frequency as the motor turns faster. Electrons also 'hum', but their vibrations are in the higher frequencies of the electromagnetic register – visible light, X-rays, …… When examined as in slow motion, this emitted radiation is no longer a continuous hum: instead it is a rapid staccato of radiation bullets, or quanta, as a tumult of applause is built up of individual handclaps. Einstein had called these radiation quanta 'photons', to rhyme with electrons and underline the particle-like behaviour of electromagnetic radiation.

An electron signals its presence like a quantum radar beacon, continually emitting and absorbing photons. These transient photons surround the electron like an electromagnetic halo, its field. When another electron approaches, some photons, instead of returning to their home electron, become entangled instead in the halo of the second electron, and transfer their allegiance. Thus is the influence of the electromagnetic force transmitted, and one electron 'feels' the presence of another.

To describe such effects, Dirac's designer equation had to be harnessed to the electromagnetic field. Grafting the equation onto the mathematics of quantum electromagnetism without losing its elegance was difficult. Great masters – Dirac himself, Werner Heisenberg and Wolfgang Pauli – groped their way forward, and their students – Victor

Weisskopf, Rudolf Peierls, Hans Bethe, Walter Heitler.... – struggled with their allotted tasks. They were playing a new game – quantum electrodynamics – while being totally ignorant of its rules, discovering or inventing them as they went along.

One who helped was Homi Jehangir Bhabha, born in Bombay in 1909 in an aristocratic family that stressed the importance of education. The family was close to the mighty Tata empire, which had pioneered India's effort in iron and steel, and in power generation, setting it on the path to industrial might. In 1927, Bhabha went to Cambridge to study engineering, but while an undergraduate, like Abdus Salam later, came under the spell of Paul Dirac, and was drawn into the world of mathematics and theoretical physics. After his degree in 1930, Bhabha began research, and in 1935 was the first to calculate how an electron and its antiparticle (a positron) annihilate into a single photon, which then reincarnates an electron–positron pair. Electron–positron interactions have been known ever since as 'Bhabha scattering'. (In the arcane quantum electrodynamics of the 1930s, such calculations were heroic. Now they are assigned as routine homework or examination problems.)

In this emergent quantum electrodynamics, one feature was crucial. Even the most complete theories ultimately depend on numbers that can only be measured by experiment. Einstein's theory of relativity said that the velocity of light is the same everywhere, but cannot predict its actual value. This key parameter can only be measured. Quantum electrodynamics has its characteristic input number, and it is very small. Assemble enough atoms together and their coherent magnetism can lift a wrecked car, but when a single photon messenger is absorbed or emitted, the effect is tiny, an effect of less than one per cent in the quantum calculations. Further photons contribute less than one per cent of one per cent, so the electromagnetic experience of a single electron should be represented fairly accurately by a token photon, as Bhabha had done.

Extending the calculations to include additional effects should have given tiny corrections, fine tuning almost too small to measure, but instead, the mathematics unexpectedly exploded. Rather than giving tiny numbers, it spat out infinities, which jammed up any chance of making calculations. Quantum electrodynamics appeared to have gone completely out of control. Dirac, rarely at a loss for an imaginative suggestion, suspected that this was could be due to the definition of the electric charge carried by a single electron. It was this charge that ultimately governed the strength of any electromagnetic effect. An electron is attributed with a specific electric charge, a tiny but fixed fraction of the amount of electricity carried by a current of one ampere during one second, but any single quantum electron is shrouded by its quantum halo, which itself mimics electric charge, masking the charge on the 'bare' electron inside.

Calculations should separate this bare charge from the extra contributions due to the electron–photon halo, alleged to be infinite. The trick was to attribute the naked electron with an additional infinite charge, the latter exactly cancelling the infinite contribution from the calculation, giving a result that agreed with the measured charge on the electron. Subtracting two infinities to give a precise result looked very suspect. There is an infinite number of numbers: add up these numbers and the result is also infinite, but more so. How can the number of numbers equal their sum? But physicists said that the infinities of quantum electrodynamics could cancel each other out. Such a trick hinted that the theory was somehow deficient, but nevertheless it worked. Quantum theorists called their mysterious procedure 'renormalization' and marched on. The trick could be cleverly camouflaged because its infinities occur only with quantities, like the charge on the electron, or its mass, which no theory can predict and have to be determined by measurement.

One loud protest came from Dirac himself, who said in the final paragraphs of his book *The principles of quantum mechanics*:

"People have succeeded in setting up certain rules that enable one to discard the infinities... in a self-consistent way and have thus obtained a workable theory. Good agreement with experiment has been found, showing that there is some validity in the rules. But the rules are only applicable to special problems, ... and do not fit in with the logical foundations of quantum mechanics. It would seem that we have followed as far as possible the path of logical development of the ideas of quantum mechanics as they are at present understood.[1]

Although Dirac was now out of the game, his equation remained the engine that powered quantum electrodynamics, until the peace that followed the Second World War cast an unexpected new shadow. Scientists who had perfected their microwave skills while developing radar technology in the Second World War returned to research at Columbia University, New York, and began to make new ultra precision measurements of the light from hot hydrogen atoms. Each type of atom, when heated, gives out light that, when passed through a prism, splits into characteristic bands of colours, a spectral fingerprint for that atom. From the distinct bands of colour in starlight, astronomers can identify what atoms are in the distant star. Helium was discovered in this way in sunlight in 1868, before the gas had ever been found on Earth.

This spectrum of each type of atom is the result of a characteristic ladder-like pattern of energies dictated by the quantum mechanics of its electrons. An electron squats on a rung of an energy ladder until disturbed, when it leaps to another rung, emitting a photon if the rung is lower, absorbing a photon if it has to climb higher. The spectrum of the atom is the end result of all the possible quantum jumps, joining the energy dots to make a picture. However, the rungs are not arranged as a simple up-and-down ladder, but resemble instead more a scaffolding. According to the Dirac equation, two adjacent girders in the scaffold of the hydrogen atom should touch, so that electrons arriving at the same point via different energy paths would have gained or lost the same amount of energy. But measurements by Willis Lamb in 1947 revealed a tiny mismatch – the 'Lamb shift'. Two girders that were expected to touch in fact did not. For the first time, the Dirac equation had failed. The anomaly was only a few parts per million, but it had been measured. Other precision microwave experiments soon revealed more such discrepancies.

Young theorists, who only a few years before had calculated the transmission of microwaves or simulated the blast of conventional explosive needed to trigger an atomic bomb, were surveying the delapidated pre-war quantum electrodynamics that had been left by their teachers. Julian Schwinger and Richard Feynman threw away that shabbiness and fashioned a new elegance that disguised the clumsiness of renormalization and showed how small effects like the Lamb shift could arise and be calculated. Initially few could understand Schwinger and Feynman, and Schwinger and Feynman moreover could not understand each other. It fell to a young British student, Freeman Dyson, to play an analogous role for quantum electrodynamics to that Dirac had played for quantum mechanics twenty years earlier. Dyson had been lured into mathematics by becoming intrigued by Ramanujan's work in number theory, and went on to meet Godfrey Hardy at Cambridge before being recruited for the war effort. Arriving on the US research scene soon after Feynman and Schwinger, he reconciled the apparently disconnected approaches, understood the clashing infinities, and made their accounting respectable.

After carefully excising the numerical blemishes, calculations reached awesome precision – equivalent to predicting the exact number of people on the planet at any second in time. A theory that can make such predictions must be doing something right, and quantum electromagnetism became the prototype theory for others: physicists were impatient to write down analogous theories for Nature's other agents. Electromagnetism turns the cogs within the atom. Very different forces grip the proton and neutron inside the atomic nucleus. These forces have to be enormously strong to overcome the huge electrical repulsion between the positively charged protons crammed tightly together, but they are tightly confined inside the tiny nucleus at the heart of the atom. Outside the nucleus, its immense forces are invisible.

In Japan, Hideki Yukawa took 1930s electromagnetism, with its halo of photons round each electron, as a model for a nuclear picture. Yukawa's idea was that protons and neutrons are surrounded by a much more compact halo, made of lethargic heavy particles that could shuttle back and forth. These hypothetical messengers were called 'mesons', particles more massive than electrons, but lighter than protons and neutrons. In 1947, the British physicist Cecil Powell took slabs of photographic emulsion to the tops of high mountains and recorded the tiny traces left by cosmic rain, subnuclear droplets from outer space, before they could be soaked up by the atmosphere lower down. Examined through a microscope, his photographs showed how these cosmic droplets shattered nuclei, producing unstable fragments whose life history had been imprinted in the photographic emulsion. Powell had seen the mesons of Yukawa and was moved to poetry: 'It was as if, suddenly, we had broken into a walled orchard, where protected trees flourished and all kinds of exotic fruits ripened in great profusion.'[2] The hour had come to write down the quantum dynamics of these mesons, mimicking the successful picture of electrodynamics. If calculations were to be possible, any infinities thrown up by meson theory had to be brought under control: meson dynamics had to be renormalizable.

Paul Taunton Matthews was to play an important role in Salam's life. 'My best friend', said Salam much later[3]. Like Salam, Matthews was born in British India – in Erode, Madras, in 1919, where his parents were missionaries, his father teaching English at Madras Christian

College. Although Matthews left India at the age of seven to attend school in England, his childhood memories of the subcontinent must have added to his bonding with Salam. Strongly principled, Matthews registered as a conscientious objector in 1940, and spent the war as an ambulance driver, first in London, and then in China, overseeing the distribution of medical supplies. Before the war intervened, he had been a wrangler in Part II mathematics at Cambridge, and after the war returned to attack the challenge of Mathematics Part III. Here he heard the call of Paul Dirac, as had Bhabha, as did Freeman Dyson, prior to his departure to the United States, and as would Abdus Salam[4]. In 1947 Matthews married Margit Zohn, from an Austrian Jewish family who had fled from Vienna in 1939, and began research in theoretical physics at Cambridge under the supervision of a man who oversaw a generation of talented post-war graduate students at a time when theoretical subnuclear physics was making spectacular progress. That man was Nicholas Kemmer.

Born in St. Petersburg in 1911, Kemmer came to London in 1916 when his father's career had taken a fresh turn before the 1917 October Revolution. In 1922, the family moved again to Germany, where Kemmer later became a student at the University of Göttingen, at the time the world centre of quantum theory, and was taught by quantum pioneer Max Born. (After the Nazis came to power in 1933, Jews were sacked from public positions, including university posts. Born moved to Britain, and in 1936 was appointed to the Tait Chair of Natural Philosophy at Edinburgh. He was awarded the Nobel Prize for Physics in 1954. Fate eventually linked the career paths of Born, Kemmer and Salam.) Kemmer spoke several languages with impeccable accents, making it difficult to tell what his mother tongue had been. Transferring to Zurich to do research in theoretical physics, in 1932 he became an assistant to Wolfgang Pauli. In 1936 he moved to Imperial College London, where he elaborated Yukawa's theory of subnuclear forces, correctly predicting that the mesons that carried these forces would have to come in three electric charge versions; positive, neutral and negative. After wartime nuclear fission work, in 1946 Kemmer returned to Cambridge, soon to become Stokes Lecturer in Mathematics.

Under Kemmer, Matthews assessed the renormalizability of meson interactions, extending to subnuclear physics what Dyson had done for electrodynamics. If meson theory were to be renormalizable, Matthews showed what kinds of mesons were possible and how their theories could be constructed. In 1949 he was writing up this work for his doctoral thesis before becoming a Visiting Research Fellow at the Institute of Advanced Study, Princeton, the ivory tower where Einstein still resided, but no longer reigned: the new quantum physics had become the business of younger minds.

While Matthews was busy writing his thesis, Abdus Salam was a research novice, blundering about in the Cavendish Laboratory, scene of his earlier disastrous encounters with the art of experimental physics. Now he had been instructed to measure what happened when nuclei of heavy hydrogen crashed into each other. During Rutherford's reign at the Cavendish, nuclei had been analysed into protons and neutrons. A single proton was the nucleus of the lightest element, hydrogen, but there were also exotic forms of hydrogen with one or two neutrons grafted onto the trademark proton.

Experimental work did not go well for the young research student. 'Soon,' Salam admitted in his Nobel Prize speech in 1979, 'I knew the craft of experimental physics was beyond me – it was the sublime quality of patience, patience in accumulating data, patience with recalcitrant equipment – which I lacked'[5]. So he despaired of the laboratory, leaving experiments on heavy hydrogen to others with more patience, and instead began to study quantum field theory.

'What are you reading?' asked a friendly Paul Matthews at their first encounter. 'Heitler,' replied Salam. Walter Heitler's *The quantum theory of radiation* had been published in 1936, when renormalization still implied confusion, and before the Lamb shift had been discovered. 'Reading' such a textbook was not a matter of turning pages. Every page was packed with equations, each of which had to be worked out afresh if the student could progress to the next. Salam was slowly ploughing his way through the book. Matthews pointed out that great things had just happened that had yet to hit the pages of books. Matthews told Salam to forget Heitler and follow instead the equations of Richard Feynman and Julian Schwinger published in the pages of the US journal *Physical Review*. Salam complied. He had already found that textbooks quickly became out-of-date in fast-moving modern research. He had bought a copy of a classic but very expensive book on nuclear physics just before listening to a lecture by Hans Bethe, who was passing through Cambridge. A refugee from Nazi Germany, Bethe was the first to understand how the Sun is fuelled by thermonuclear fusion, for which he later received the Nobel Prize for Physics in 1967. During the war, he had been head of the Theory Division at the Los Alamos atomic bomb laboratory, and had also discovered and nurtured the young Richard Feynman. Listening to Bethe's survey of the nuclear physics frontier, Salam realized his new book was out-of-date, and rushed back to the university bookshop, demanding his money back. However, the book was now deemed to be second-hand and worth only a fraction of the price Salam had paid for it. After that, Salam distrusted textbooks. They were fine for undergraduates, but got out of date too quickly to be useful in frontier research.

Now better informed, in December 1949 Salam went to see Kemmer and asked to be taken on as a research student in theoretical physics. He did not know that Kemmer was already trying to resist pressure from his peers to take him on as another research student[6] – Salam's examination performance could not be ignored. But with his hands already full with eight other research students, Kemmer did not want any more. He did not expect any newcomer to be as easy to manage as Paul Matthews had been. When he eventually met Salam, Kemmer was still not impressed by the subservient young applicant ('I nearly refused Salam', he said later[7]), and suggested that he went instead to Rudolf Peierls in Birmingham. Some ten years before, Peierls had guided Fred Hoyle's first steps in research at Cambridge, and had later played a key role in the development of the wartime atomic bomb, where he had been among the first to realize just how compact a critical mass of fissionable nuclear matter could be. Salam had been living in Britain for more than three years, but was uneasy about moving to a strange, large city. He knew Cambridge well, and felt comfortable in its great machine of learning. He wanted to do research and live in college, not to have to fend for himself in a place he did not know. In Pakistan, his wife was now expecting a child. Above all, he was confused and depressed after his fruitless tryst with experimental work, his sudden plunge into deep theoretical waters, and the cool reception from Kemmer. After having followed the advice of his colleagues who had told him to move into research, Salam was now angry and frustrated. His Indian contemporary Ram Prakash Bambah recalls Salam alleging that they had 'misguided' him, and using 'very strong Punjabi expressions' in his disappointment[8]. Salam pleaded with the haughty Kemmer, asking to be taken on 'peripherally', and this time was told to go and talk to Matthews.

So began what was to become Salam's most important scientific collaboration, and a lasting friendship. Some scientists do their best work in solitude – Dirac was a good example. Salam was otherwise. He was to become a highly imaginative scientist, sometimes almost too imaginative, and learned that he functioned best when he worked with a partner with whom he could argue out his ideas and who could channel his inventiveness. The first was Paul Matthews, who had an uncanny sense of research trends, sensing the direction in which the wind of discovery would blow, and always had an up-to-date knowledge of all the latest subnuclear particles to have been discovered in Powell's 'walled orchard'[9].

When Salam came begging for research crumbs, Matthews immediately told him to move on from the 1947 papers of Feynman and Schwinger and look at what Freeman Dyson had done for

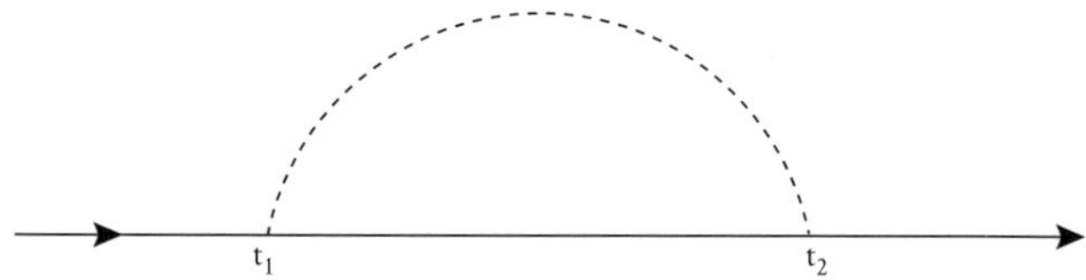

Diagram 1 The timeline of an electron, showing how, at time t_1, it 'decides' to hurl out an electromagnetic photon (dashed line), with scant respect for conservation of energy and momentum. To qualify for such dispensation, it reabsorbs the photon at time t_2, within an interlude given by quantum uncertainty rules. As electrons do this all the time, such effects have to be allowed for. Quantum pioneers had to master the awkward infinities thrown up by these calculations.

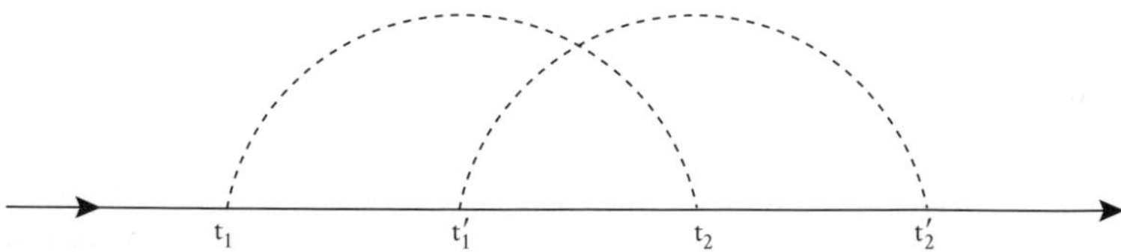

Diagram 2 Sometimes an electron can hurl out another photon (at time t_1') before the preceding one has returned to base. Theoreticians had swept such troublesome problems of 'overlapping infinities' under the carpet until Salam showed in 1950 how they could be handled. It was his first contribution to research.

renormalization. In Matthews' renormalization scheme for mesons, he had concentrated on 'one-loop' processes, allowing only one meson at a time (Diagram 1). He suggested to Salam to take the next step, where any number of mesons could enter the calculation. Salam rushed back to his papers, reread what Dyson had done, and in a few days triumphantly dashed back to Matthews, brandishing his 'solution'. Matthews laughed. 'You've [shown] the various factors do fit and everything would be fine – if one could show that the infinities really can be consistently removed. That is the problem.'[10] It boiled down to 'overlapping divergencies', when a second meson is emitted before the first has found its destination (Diagram 2). In such complicated tangles of meson interactions, could each infinity be surgically removed as if the others were not there?

Everyone assumed that Dyson had shown how this could be done for electrodynamics. In the spring of 1950 Matthews had his PhD *viva* examination with Dyson, who worked most of the time in the US, but came to Birmingham for a few months each year. In the interview, Dyson suddenly asked Matthews how he was taking care of overlapping divergencies in meson theory. Surprised, Matthews claimed that Dyson had assured everyone that such processes would pose no problem, and had taken this for granted. No further question was asked, but Matthews remained uneasy. In the electrodynamics that Dyson had studied, such enigmatic overlapping divergencies were rare, but Matthews' subnuclear processes had a rash of them. With his PhD accomplished, he let go of the nagging problem and announced to Salam that he was going on vacation before moving to the United States. Their agreement was that Salam could look at the problem over the 1950 summer vacation and make whatever progress he could. When Matthews returned from his vacation, he would take back the problem for himself and resume his quest to renormalize meson theory.

Salam had been given a hard nut to crack, and had little time. He knew that the person who understood renormalization better than anyone else in the world was Dyson, who at that moment was still in Birmingham. Salam made a phone call. 'I am beginning as a research student,' he explained, 'there is this problem of overlapping divergencies'. Dyson could have been irritated by a call about a worrisome problem from someone he did not know. He was also preparing to leave for the USA the next day. 'You must come immediately,' he said, maybe thinking that was the end of the matter. Undeterred, Salam jumped on a train to Birmingham.

Dyson's classic paper on renormalization had appeared in *Physical Review* in 1949. At Birmingham, Salam pleaded with Dyson for a clue to the rebellious mathematical terms that were blocking the route for meson theory. But Dyson admitted that while he had shown how other rebellious terms could be brought into line, he had simply assumed that the overlapping divergencies would follow suit. Salam was shocked. He now saw the size of the obstacle Matthews had been up against. Dyson's renormalization scheme was not mathematically rigorous and contained a conjecture. But it worked. Its value lay in its usefulness, not yet in its mathematical rigour. On that spring day in Birmingham, Salam saw how speculation and conjecture drove research forward. Anxious not to let go of his fragile hold on the problem, he stayed overnight in Birmingham with physicist Richard Dalitz, who he knew from Cambridge. The next day the insistent Salam dogged Dyson, *en route* for the United States, sitting with him on the train to London. During the two-hour journey, Dyson confided some private ideas that to him justified accepting the overlapping divergencies. As Salam travelled back to Cambridge, a conceptual crack widened and he began to see how he could vindicate what Dyson had glossed over. That summer, walking in the rose gardens of St. John's, Salam saw what had to be done, and in the evenings began to write it down.

As with many problems in mathematics, the trick was to think of the right framework. Salam transferred the mathematics from the four-dimensional space-time of Einsteinian relativity into the less familiar territory of a four-dimensional momentum space, where it was easier to classify the unruly infinities and finally bring them under control. Returning from vacation, Matthews was amazed to find Salam now had a complete solution. The end product, a ten-page paper 'Overlapping Divergencies and the S-Matrix' arrived on the editorial desk of *Physical Review* in New York on 29 September 1950. The first line refers to Dyson's milestone paper, and goes on to say exactly what Dyson had showed, and what he had not showed. The text is spattered with blocks of impenetrable mathematics and with diagrams that resemble simple crystalline structure but that in fact demonstrate how particles can emit and absorb mesons. The final paragraph says that the author was 'deeply indebted' to Dyson for 'an extremely helpful discussion, without which this work would not have been possible'. Salam also acknowledges the remote figure of Kemmer for 'continual help and encouragement'.

The paper did not appear in print until April 1951, but with such fast-moving research, important results were increasingly being issued as 'preprints', duplicated typescripts distributed by airmail to libraries in major research centres, there to be pounced on by eager scientists. Through this preprint channel, the name of Abdus Salam quickly became known. This paper was a research milestone, and alone could have qualified him for his doctorate, but Cambridge University regulations stipulated that he could not do this until 1952, after three years as a research student. Matthews had left for a research fellowship at Princeton. Nicholas Kemmer now realized the extent of the talent he had nearly overlooked, and pointed Salam too towards Princeton, where as a Visiting Research Fellow he could continue his collaboration with Matthews. However, as a Cambridge research student, Salam needed to have a formal research 'supervisor' while working elsewhere. Matthews took on that role. Their paper also merited Cambridge University's prestigious Smith's Prize, dating from 1768 and awarded annually to research students in theoretical physics, mathematics and applied mathematics deemed to have made the greatest progress in mathematics and natural philosophy[11].

There is a time in the life of a scientific genius when a crushing burden of responsibility is felt for the first time, when ability has to be channelled into discovery. Salam now felt this pressure, and learned that he could best strive to make discoveries with a partner, and Paul Matthews was the perfect one. Later, Salam would seek others to play this role. Another genius of the twentieth century, Wolfgang Pauli, worked throughout his research life with a series of research assistants, many of whom went on to become famous scientists in their own right. Part of Pauli's job description for these assistants was 'Every time I say something, contradict me with detailed arguments'[12] Salam too needed captious accomplices.

Subnuclear physics had been born in Europe, but the effort to develop the wartime atomic bomb had swung the intellectual pendulum westwards across the Atlantic. Earlier in the twentieth century, young American scientists had learnt in European, but now they had snatched the baton and were running ahead. Amid America's new intellectual vigour, Princeton's Institute for Advanced Study rapidly emerged as one of the acknowledged world centres. Unlike Europe's prestigious universities or the US Ivy League, it had little history and tradition: after the Bamberger family sold their New Jersey department store to Macy's of New York just before the stock market crash, they founded the Institute in 1930 'to encourage and support fundamental scholarship'. It is not a university, and there are no formal courses of study. Initially named after its first director, Abraham Flexner, in its early years it attracted intellectual refugees from Nazi Germany, and its figurehead was Albert Einstein, who was there for the remainder of his life. Other recruits included mathematicians John von Neumann, Hermann Weyl and Kurt Gödel. Wolfgang Pauli worked there during the Second World War, and the announcement that he had been awarded the 1945 Nobel Prize for Physics was a major boost for the Institute – the first time a member had been awarded a Nobel Prize while in residence.

The atmosphere at the Institute changed dramatically after the War as scientists who had toiled in wartime programmes returned to peacetime research. After having been scientific director of the US atomic bomb project at Los Alamos, in 1947 J. Robert Oppenheimer became the Institute's Director, and turned his attention from that vast wartime effort to the emerging new directions in mathematical physics. Among his early post-war recruits were Freeman Dyson, apostle of the new quantum electrodynamics, and T. D. Lee and C. N. Yang, former students of Subrahmanyan Chandrasekhar in Chicago and later to be the central figures in startling new research revelations. Another was a physicist called Murray Gell-Mann, totally unknown, but who stood out because at 21 he was much younger than everyone else. Dyson's initial objective on arrival at Princeton was to convince Oppenheimer, still entrenched in pre-war quantum theory, of the importance of the new quantum electrodynamics. Although initially wary of the unfamiliar new techniques, Oppenheimer was soon to act as a father-figure to a post-war theory boom, presiding over historic meetings and cajoling young researchers. Extending the techniques of quantum electrodynamics to meson theory and polishing its renormalization procedures were high on the list of objectives.

In January 1951, and for the second time in five years, Abdus Salam's intellectual achievements beckoned him to rush to become an inhabitant of a country that was strange to him. Arriving at New York, this time there was no Zafrullah Khan to meet him at the dockside. On his way across town to the railway station, he must have marvelled at the skyscraper canyons of Manhattan after the twisting lanes of Cambridge and the ancient temples of Lahore. The United States is vast, but Princeton is only about 100 kilometres from New York City, a short trip, even by European standards. Just as Cambridge had ensured a soft landing for him in Britain, so the monastic atmosphere of Princeton's Institute for Advanced Study shielded Salam from the brashness of American life. And the American diet was certainly more ample than that of postwar Britain. The intellectual village with its green campus and neo-classical buildings was the nearest thing in America to Cambridge. There were few distractions, 'no pubs, not even a decent coffee shop', wrote another British visitor, John Ward, later to be one of Salam's key research collaborators[13]. The Institute was a place to throw oneself into one's work.

This time, Salam was not an unknown student who could hide in the library. He was embarrassed to find that his reputation had preceded him. Despite having only one year's experience, he was greeted as a research phenomenon. Here was the man from afar who could make unwanted infinities disappear. Paul and Margit Matthews were there to greet him. There was also Freeman Dyson. Salam did not speak to Einstein, but frequently saw the great man wandering on the lawn, always wearing a hat, lost in thought. As well as a reputation, Salam also arrived with the manuscript of his next renormalization paper, but before sending it off for publication, forwarded a copy to Oppenheimer. In his haste, he was shocked to discover that he had sent his boss the wrong copy, a draft without the detailed illustrations that displayed the tortuous interactions of the particles. Rushing to retrieve the paper, he bumped into Oppenheimer, who had already read it. 'I enjoyed reading your paper. It is a fine paper,' he said. Salam related 'I should have kept quiet, but like a fool I said "I am sorry, but I gave you a copy in which there were no diagrams. I don't think you could have understood it." Oppenheimer visibly changed colour, but only murmured "The results are surely true and intelligible without diagrams".'[14]

Salam was awed by the celebrities at Princeton and felt a long way from home. 'The result was disastrous,' he wrote later[15], 'I knew nothing of physics except what I had done for myself. I was afraid to reveal my ignorance. I learnt nothing new'. However his claim was an exaggeration. Another young researcher there in 1951 was the Swiss mathematical physicist Res Jost, who had learnt his trade in Zurich during the Second World War. Before moving to Princeton, he had become one in the series of illustrious research assistants employed by Wolfgang Pauli at the Zurich Federal Technical Institute. In the mathematical treatment of his renormalization problems, Salam acknowledged help from Jost. (After leaving Princeton in 1955, Jost returned to Zurich and in 1958 eventually inherited Pauli's position.)

As a distraction from the monk-like research life at Princeton, a significant visitor was the distinguished upright figure of Zafrullah Khan, who five years before had befriended a confused Salam at the dockside in Liverpool. In the intervening time, Salam had progressed from undergraduate to research prodigy, while Zafrullah Khan had become Pakistan's delegate to the United Nations Assembly in New York. With Zafrullah Khan, Salam took time out from research to tour the East Coast of the United States, including New York, where construction work was still underway for the impressive new UN headquarters. It was Zafrullah Khan who introduced Salam to the stage of the United Nations, later to become a vital part of Salam's master plan for new scientific ventures.

Returning to Princeton, Salam threw himself into his work with Paul Matthews. Although a research fellow at the Institute for Advanced Study, Salam was still a Cambridge research student and Matthews was technically his research supervisor. But this was no ordinary teacher–student relationship, and in May, he learned that he had been elected a visiting Research Fellow at St. John's, strengthening his position. Meson theory had become a Salam–Matthews joint venture: their collaboration was highlighted at the annual jamboree of the American Physical Society, held in Schenectady on 16 June. Their joint paper was a calculational framework for subnuclear physics, the logical next step after Schwinger and Feynman's quantum electrodynamics that had dominated the American Physical Society's meetings in previous years. It looked like physics had arrived at a new threshold. Dyson had told Oppenheimer 'We have now a theory of nuclear fields which can be developed to the point where it can be compared with experiment'[16]. Salam said 'We believed this was the end. We expected it to be *the* theory. The end! And we lived in that paradise, euphoric for a year.'[17]

The *Physical Review* was the prestigious American journal in which important developments had to be duly logged. But with developments happening fast, it was good to have periodic summaries, review papers, where breathless research workers could take stock. This was the aim of the stately *Reviews of Modern Physics*. In October 1951, the journal carried 'The Renormalization of Meson Theories' by Paul Matthews and Abdus Salam. It was a written version of the talk that had been given at

Schenectady. The crude, hand-drawn sketches representing quantum particle behaviour belie their scientific importance. Here was the promise of an ultimate theory, which would explain the intricate mechanisms of subnuclear mechanics and enable precise calculations to be made, just as quantum electrodynamics had explained and gauged the Lamb shift. It looked as though the reputation of Abdus Salam and Paul Matthews would soon match those of Richard Feynman and Julian Schwinger[18]. But Salam and Matthews had been working in a institute whose research was totally theoretical: there was no adjacent laboratory. Experiments at distant laboratories had given unexpected results, and from these first enigmatic wisps, clouds began to gather on what had been a euphoric research horizon.

REFERENCES

1. Dirac, P. A. M., *The principles of quantum mechanics*, (Oxford, OUP, 1958)
2. Powell C. F., Nobel Prize lecture, 1950
3. Salam, A., *A life of physics*, in Cerderia, H. A., Lundqvist, S. O., (ed.) *Frontiers of Physics, high technology and mathematics* (Singapore, World Scientific, 1990)
4. Kibble, T. W. B., Paul Taunton Matthews, *Biographical memoirs of fellows of the Royal Society* 34, 555–80 (London, Royal Society, 1988)
5. Salam, A., *Gauge unification of fundamental forces*, Reviews of Modern Physics, 52, 525–39 (1980)
6. Kemmer N., 'The Cambridge Days', in *Abdus Salam as we know him*, ed S.M.W. Ahmad (Trieste, ICTP, 1990)
7. Kemmer N. in Kursunoglu, B., and Wigner, E. P., (ed.) *Reminiscences of a great physicist, Dirac*, (Cambridge, CUP, 1987)
8. Ram Prakash Bambah, *Together in Lahore and Cambridge*, in Hamende, A. M. (ed.), *Tribute to Abdus Salam*, (Trieste, ICTP, 1999)
9. After Powell's initial foray, the fruits in this orchard were now being cultivated by experiments at US accelerator laboratories.
10. Salam, A., *A life of physics*, in Cerderia, H. A., Lundqvist, S. O., (ed.) *Frontiers of physics, high technology and mathematics* (Singapore, World Scientific, 1990)
11. Past recipients of the Smith's prize include: 1841 – George Gabriel Stokes; 1845 – William Thomson, Baron Kelvin; 1854 – James Clerk Maxwell; 1880 – J. J. Thomson; 1901 – G. H. Hardy; 1907 – Arthur Eddington; 1936, Alan Turing; 1938, Fred Hoyle.
12. Enz, C. , *No time to be brief – a scientific biography of Wolfgang Pauli*, (Oxford, OUP, 2002) 195
13. Ward, J. C., *Memoirs of a theoretical physicist* http:/www.opticsjournal.com/jcward.pdf

14. Salam, A., *Physics and the excellences of the life it brings*, in *Pions to quarks*, Brown, L., Dresden, M., Hoddeson, L., (ed.) (Cambridge, CUP, 1989).
15. Salam, A., *Physics and the excellences of the life it brings*
16. Gleick, J., *Genius, Richard Feynman and modern physics*, (New York, NY, Little. Brown, 1992)
17. Salam, A., *A life of physics*, in Cerderia, H. A., Lundqvist, S. O., (ed.)
18. The new developments of quantum electrodynamics had been developed in parallel by Sin-Itiro Tomonaga in Japan. In 1965, Tomonaga, Schwinger and Feynman shared the Nobel Physics Prize.

7

Not so splendid isolation

The hopes of a workable subnuclear field theory did not last long. Physicists had been seduced by the benevolence of electromagnetism, its small effects making approximate calculations so easy, like borrowing money at an annual interest of less than one per cent. Soon new experiments showed that subnuclear forces were less tractable. Although they only acted over tiny distances, much smaller than an atom, the number that fixed the scale of these forces was some 2000 times greater than that of electromagnetism. Any loan at such an exorbitant rate of interest would be quickly swallowed up by the accrued debt. The calculational framework so carefully constructed for electromagnetism and that gave such impressively precise results came crashing down in the subnuclear arena. Even if infinities were carefully removed by renormalized accounting, the surrounding calculations became meaningless. At the end of their 1951 paper in *Reviews of Modern Physics*[1], Matthews and Salam valiantly tried to finish on an upbeat note, noting that the general structure of their theory agreed with the observed characteristics of mesons. But they added glumly 'Beyond this very little can be got from the theory in its present form.'

Soon, other subnuclear effects were discovered that further undermined any hope. Powell's 'walled orchard' had initially revealed several kinds of subnuclear fruit, mesons whose exact role was not entirely clear, but nevertheless hinted at new understanding. Instead of waiting patiently on the tops of mountains for such sporadic mesons to descend from the sky, restless physicists wanted instead to beat Nature. In the United States, nuclear science had played a major role in the war and its scientists emerged as heroes. To preserve the strategic importance of this science, the nation endowed it with huge new laboratories, on a scale to match what had been done at Los Alamos. These were equipped with particle accelerators – a new kind of machine, related to the vast electromagnetic centrifuges that had winnowed the uranium isotopes for the Hiroshima bomb. Inside these particle accelerators, ribbons of particles whirled round a magnetic racetrack, faster and faster, before finally being peeled off and smashed into a target. Unravelling these complex collisions revealed more subnuclear exotics, and Powell's orchard was soon looking more like a jungle. Within a few years, a whole forest of unstable subnuclear particles had been uncovered, each of which lived for a brief fraction of a second before decaying in a tiny firework display. A few particles lighter than protons and neutrons could be understood as the ligature that held them together in the nucleus. Now there were particles *heavier* than protons and neutrons. There was no script for all these players that had burst uninvited onto the scene. Faced with such dilemmas, the theoretical emphasis turned from quantitative to qualitative understanding: what was this extraneous subnuclear stuff? Quantum field theory, essentially quantitative and that had been riding the crest of a wave since the discovery and explanation of the Lamb shift, suddenly went out of fashion. Field theorists like Salam were easily wrong-footed in the ensuing research scramble.

At this point, Salam's scholarship, originally from British India but taken over by the government of the new nation of Pakistan, expired. Salam had climbed the highest summits of research, but had done so faster than the Cambridge PhD regulations allowed. He therefore had yet no doctorate, the conventional entry ticket to academia, but did have an offer of a temporary position at the Institute for Advanced Study at Princeton, where important work was being done. Standing at a crossroads after five years away from home, Salam felt his duty was to return to the Punjab, now part of a nation he had not yet really experienced. Nor had he experienced his new family: his daughter, Aziza, had been born in Multan in June 1950, while he was at Princeton.

Salam saw his daughter for the first time in the autumn of 1951 when he returned to the Punjab to become Professor of Mathematics at Government College, Lahore, his alma mater, and Chairman of the Mathematics Department of the University of the Punjab. Here he could be a new Chowla, imparting the rigour and mystery of mathematics to new generations of students. His undergraduate teaching was at Government College, while graduate instruction was organized by the University, where students from different colleges came together. Mathematics is a key discipline – all nations need scientists and technologists, and all scientists and technologists need mathematical training. In this way, Salam could help his new country struggle to its feet after a troubled birth. As part of the applied mathematics curriculum, Salam initially taught electricity and magnetism – fitting for one who as a boy from a remote town had had his first encounter with electricity when he came to Government College as a student a decade earlier.

The syllabus in Lahore had yet to feel the full impact of twentieth-century scientific advance: quantum mechanics was still considered outlandish, rumour rather than knowledge. To sidestep this inertia, Salam suggested an evening course in modern quantum theory, outside the standard curriculum. After about three lectures, only two students remained, the twin brothers Riazuddin and Fayyazuddin. With such a dismal attendance, Salam stopped giving the course, but he had been impressed by the enthusiasm and talent of the brothers. Fayyazuddin was doing a masters degree in physics, where he had been assigned the problem of measuring the energy loss of cosmic ray particles as they penetrated the atmosphere. Salam told him that this was the wrong way to look at the problem: rather than simply measuring the energy loss, it would be better to try to search for a pattern of behaviour, for example seeing how the energy loss changes with the energy of the incoming particles. The suggestion was fruitful, and Fayyazuddin's physics teacher was impressed. Riazuddin preferred mathematics, and went on to become Lecturer at Government College after completing his MA in 1953. When Salam eventually left Lahore, Riazuddin took over his teaching. However, not all the students appreciated Salam's innovation. Many had found his unconventional examination questions difficult, full of unfamiliar quantum paradoxes.

In Cambridge and Princeton, Salam had learned how to construct new theories, and hoped that he would be able to build fresh ones in Pakistan. It needed talent, which he had, and a supply of paper and pencils from Government College. But he soon discovered that it also needed something else, which Pakistan could not supply. With quantum physics not even on the student curriculum in Lahore, and Salam's bid to introduce it through extracurricular lectures initially a failure, his fragile platform for modern field theory was being steadily undermined. After the heady intellectual atmosphere of Princeton, working alongside the best minds in the world, and with a hot line to who was discovering what, even his return to Cambridge and the rose gardens of St. John's had been something of an anticlimax. Cambridge was still one of the world's premier universities, but it was a lot further away from the subnuclear action than it had been in the days of Rutherford.

Lahore was a world away from Cambridge, virtually untouched by twentieth-century science.

Lahore was also a city changed since Salam had last lived there as a student in 1946. He had not seen the mass migrations and the accompanying terror and slaughter of 1947, and had to grope his way forward in a new country where few people knew whatever flimsy rules that applied. Salam had a reputation, enough to earn seven years' immediate seniority on appointment, but as yet no PhD. Some saw a young upstart, too big for his boots, a high-flying student who had escaped the double trauma of the partition of a country and a province. Sirajuddin, the Principal of Government College, told him sternly to forget about the research work he had done in Cambridge and Princeton. The Principal, with a degree in English from Oxford, wanted good college people, stressing that Salam's main responsibility was to teach. Salam accepted this, but expected also to be able to continue his research as part of his job. However, neither the College nor the University subscribed to research journals for him to follow what was going on elsewhere in the world. Salam's annual salary was enough to live on, but not enough to cover personal subscriptions to *Physical Review*. In addition to teaching, he was expected to be a good 'college person', taking on extracurricular responsibilities. Sirajuddin offered him three alternatives: looking after the college finances; being warden of a hall of residence; or managing the college's soccer team. Salam, never a sportsman, ironically chose the latter. The aimless gesture matched his own disappointment.

Salam was to have more run-ins with Sirajuddin[2], who earlier had vied with Salam for the attention of Urmilla, the elder daughter of the former Principal of the College. Sirajuddin's 1951 confidential report alleged 'Salam is not fit for Government College, Lahore. He may be excellent for research, but not a good college man.'[3] This initially prevented him from having a college residence, and instead he rented a room in the house of Qazi Muhammad Aslam, a former Cambridge man and Professor of Psychology and Vice Principal of Government College. Abdul Hamid from the mathematics department was also helpful. While he tried to find his feet in Lahore, Salam's wife and daughter remained in Multan. Throwing himself instead into his work, he tried to ignore the cold wind of intellectual isolation. While still at Princeton, he had written more papers on renormalization with Matthews that were still in the mill. When they finally appeared later in 1951 and in 1952, Salam was listed as working at Princeton's Institute for Advanced Study, but an asterisk pointed to a footnote: 'Now at Government College, Lahore, Pakistan'.

Sirajuddin had other grudges against the polyvalent Salam, who earlier had outshone many of Sirajuddin's contemporaries as an English student. He deemed that Salam was not a team player, despite his new responsibilities for the college soccer club. Like many genius figures, Salam was not a gifted teacher of the masses, his mind continually exploring the subject rather than explaining it. He expected students to make the effort to step up, rather than him condescending to step down. Salam could not emulate the way Chowla had motivated his students and get them to attack Ramanujan problems. Despite his successes with high-flying students like Fayyazuddin and Riazuddin, the chasm between him and the mass of undergraduates underlined his feeling of isolation.

In 1952 Salam persuaded the Education Department of the Punjab Government to give him travel money, and reappeared that summer at St. John's College, Cambridge, where he was now qualified to reside as a college fellow with a meagre allowance. Dirac and Kemmer had given their support. Here he wrote another paper with Matthews, this time using the mathematics of integral equations (Fredholm Theory). This was connoisseur field theory, but with dark clouds overhanging the whole subject, the outcome moved no nearer to any physics objective. Matthews soon moved from Cambridge to work under Rudolf Peierls at the University of Birmingham, where Salam was a visitor in the following summer, with support this time from the British Council, and when more renormalization formalism ensued[4]. His return to the UK provided finally the opportunity to complete his Cambridge PhD formalities, and he defended his thesis under scrutiny by Peierls. One probing question left Salam with an uneasy feeling, and haunted him for the next few years. The neutrino, then still a hypothetical hallmark of nuclear beta decay, was supposed to have no mass. 'Why?' asked Peierls. The very masslessness of the photon, the carrier of electromagnetism, made that theory work. Why should another particle also be massless? Did the neutrino hold a message for physics? The question nagged at the back of Salam's mind.

Back in Lahore, in November 1952 Salam got word that Wolfgang Pauli would be stopping at the Institute for Fundamental Research in Bombay. The centre had been established in 1945 by Homi Bhabha and funded by the mighty industrial empire founded in the nineteenth century by Jamsetji Tata. Pauli was taking advantage of a long-standing invitation from Bhabha to visit India. After a wartime residence at the Institute for Advanced Study, Princeton, Pauli had subsequently been a frequent visitor there, but did not overlap with Salam. He had first met Salam in Zurich, where Freeman Dyson, urged by Salam, had introduced them during a physics meeting in 1951. Salam had the draft of a new paper on renormalization, which he asked Pauli to read. The quantum master, pressed for time and confronted by a pushy student he did not know, refused, with the excuse that his eyes were bad. Salam did not press Pauli any further on that occasion, but Pauli soon learned what Salam had accomplished.

As the world's oracle of field theory, all its new developments came to Pauli's attention, and he had been in correspondence with Matthews[5]. During his Indian trip, Pauli had more on his mind than sightseeing, and wanted to talk about field theory, so cabled the nearest expert – Salam in Lahore. Salam, eager to meet the oracle, cajoled an air ticket from the Bombay Institute. Pauli's opening words to the weary overnight traveller were 'The problem is, if we have derivative terms in Schwinger's action principle...'[6]. In the course of the next few years, Pauli was to hear a lot more from and about the young Pakistani physicist. On return to Lahore, Salam was immediately reprehended by the truculent Sirajuddin for unauthorized absence.

(Another by-product of Salam's unauthorized trip to Bombay was a collaboration with the Indian physicist K. S. Singwi on the nascent theory of superconductivity, the phenomenon in which certain materials, when cooled to near absolute zero, suddenly lose their electrical resistance. Although known since 1911, this phenomenon was not finally understood until 1957. Superconductivity is the result of quantum effects inhabiting bulk matter – there are no lone superconducting electrons. Physicists had begun to suspect that the new techniques of quantum field theory could help reveal the explanation for superconductivity. Salam's appearance on the scene in 1952 was typically premature, and his brief Indian collaboration resulted in his only publication in a Japanese scientific journal[7].)

As Salam's research stalled in Lahore, he became increasingly frustrated. An intensely religious man, he saw in physics the revelation of Allah's design. Several times in his talks he illustrated this with words from the Holy Qur'an: 'You can see no flaw, no incongruity, no imperfection in the creation of the All-Merciful. They look up once more: do you see any flaw? Look again and yet again, and your gaze comes back, dazzled.' (Sura 67, v3–4). Trying to probe the deepest levels of Nature's working evokes such strong emotions of awe and mystery. Albert Einstein had also felt them. A Jew by birth who during his childhood had distanced himself from the traditions of orthodox religion, Einstein had replied to a letter he received from a young girl: 'Everyone seriously engaged in science reaches the conviction that the laws of nature manifest a spirit which is vastly superior to Man, and before which we . . . must humbly bow.' [8]

Salam's introspection did not translate into aloofness. He lived his life in a continual state of ebullition that raised the intellectual temperature around him, and affected most of the people with whom he came in contact. Speaking at a memorial meeting for Salam in 1997, the distinguished British physicist John Ziman related his first encounter with Salam. Ziman had been moving from Oxford to Cambridge, and a colleague had advised 'If you are going to Cambridge, you will meet Abdus Salam. You will know who he is by his conversation', and then proceeded to utter physics buzzwords interjected by loud guffaws. Sure enough, when Ziman arrived at Cambridge, he was walking down King's Parade when he bumped into a physicist he knew, John Ward (of whom more later), with Salam, then unknown to Ziman. When conversation between the three began, Salam's speech was exactly as Ziman had been told to expect[9]. Salam talked physics the way other people might tell a joke or describe a football match. Wherever he was working, in Cambridge, London or Trieste, it was always easy to monitor when he arrived for work. He would invariably greet his working colleagues with a hearty professional joke, at which he was the first to chuckle. He had a soft, husky voice, except when he was trying to develop an idea. An intense research collaboration discussion could have been mistaken for an argument. This was his way of doing things. For Salam, physics was sheer delight.

But in Lahore his light of delight temporarily went out. In the early twentieth century, scientific research had not required huge resources. Some of the historic experiments that changed our world-view used only rudimentary apparatus: Rutherford's investigations are a good example; Raman won a Nobel Prize for laboratory work carried out in India. Nevertheless, contemporary science in the subcontinent, and in other developing countries, was increasingly handicapped by lack of special materials, skills, and above all, information. University library shelves had to be stocked with the latest scientific journals. At a time when all communication was via the printed word, scientists needed these journals to follow new research trends. But the few journals published in the West that did arrive took months to arrive by sea, by which time the science had usually moved on. The only copies of *Physical Review* in Lahore dated from before the Second World War. Salam's research lifeline was airmail letters to colleagues in England and the United States.

As well as being isolated from the engine of science, and having to contend with the jealousy of some of his colleagues, real danger began to loom. After the bloodshed of partition, the Pakistani Punjab remained at flashpoint. With the Sikhs departed eastwards for the Indian Punjab, smaller minorities became more visible, and smouldering religious quarrels burst into flame. Islamic nationalists, in opposition to the more moderate Muslim League under Pakistan's founder Jinnah, brandished the flag of religious orthodoxy. In a sense these revivalists were frustrated by the creation of Pakistan as a Muslim homeland, for power had passed into the hands of statesmen less concerned with religious matters, and who lived in Karachi, a long way from the Punjab. With Hindus and Sikhs gone, right-wing Punjabi elements quickly singled out fresh targets. Salam's family were staunch Ahmadi Muslims, a minority nowhere near as numerous as the Sikhs had been, but the several million of them in Pakistan were noticeable. In the streets and bazaars of Lahore, anti-Ahmadi feeling was stoked. Unlike turbaned Sikhs, Ahmadis are not immediately identifiable, but their places of worship are. Like anyone else, they are liable to react when provoked by taunts and insults. If gangs of youths entered a neighbourhood and started insulting the honour of Mirza Ghulam Ahmad, Ahmadi reaction soon became visible. After riots began in February 1953, the Punjab was soon locked in a minor civil war in which Ahmadis feared for their lives. The politically ambitious head of the Muslim League in Punjab, Mumtaz Daultana, used anti-Ahmadi feeling as a vehicle to further his own interests, much as Ali Bhutto was to do twenty years later, and did nothing to intervene. Transport was paralysed and the whole province ground to a standstill as Ahmadis, real or presumed, were beaten up and murdered, and their houses sacked and burnt. When the full extent of the bloodshed became evident, national government intervened and imposed martial law in the Punjab, and the provincial government tottered. Later, when the dust began to settle, an official enquiry criticized religious extremist elements. A court martial sentenced the main instigator of the troubles, the religious leader Maulana Maudoodi, to death, but the sentence was commuted to imprisonment. The episode left a livid scar on the short history of Pakistan, while Ahmadis remained isolated and vulnerable.

Salam, as Ahmadi head of a university department in Lahore, was certainly visible during these troubles. There were rumours that he had been a victim of mob violence, and to prevent the rumours from becoming fact, Qazi Muhammad Aslam, a fellow Ahmadi, smuggled Salam and his family from the spacious university bungalow that had finally been allocated to them into a safe house, away from the mobs. 'I saw scenes that I would never forget,' Salam said later, 'corpses, houses burned down, all because of my Muslim compatriots'[10].

During his summer visits to the UK in 1952 and 1953, Salam's physics colleagues asked him about life in Pakistan, and Salam told them. Already the possibility of spending a few months each year in Britain using his St. John's fellowship money to do constructive research was a major advantage. Peierls at Birmingham had suggested a prolonged fellowship, but Salam now had a family to support and needed a full-time university post. He had been hunted by rioters and had temporarily gone into hiding. He was intellectually starved and suffered the research equivalent of writers' block. Later, Salam said 'If at that time someone would have said to me "We shall give you the opportunity to travel every year to an active centre in Europe or the United States to work with your peers for three months. Would you be happy to spend the remaining nine months in Lahore?" I would have said yes.'[11] But nobody did. Even if they had, Salam would have liked an assurance that Ahmadis were no longer going to be persecuted.

Earlier in the twentieth century, the few intellectual immigrants from the Indian subcontinent had encountered obstacles at Cambridge. Even when pushed by Godfrey Hardy, an initial bid to install Ramanujan as a fellow at Trinity College in 1917 had failed, and only succeeded later after he had become a Fellow of the more prestigious Royal Society. A subsequent Trinity fellow was Subrahmanyan Chandrasekhar, who after being humiliated by Eddington in 1934 and returning to India, eventually made his career in the United States.

But attitudes in Britain changed after the Second World War. With the country no longer an imperial power, immigrants from former colonies began to be seen in a new post-imperial light. They provided an important source of fresh manpower for an economy shattered by the war. Britain was taking the first steps towards becoming a multicultural society. This was particularly noticeable in sport, seen through the prism of international competition. The Olympic Games in London in 1948 boosted the morale of a British public dispirited by their wartime experiences, bringing them a level of sporting excellence that had not been seen in Europe for a decade. Arthur Wint, who had come to Britain during the war to serve in the Royal Air Force, broke the world record for the 400 metres, giving Jamaica its first Olympic gold medal. Soon, black athletes also began representing Britain. Sprint champion McDonald Bailey, born in Trinidad in 1920, had also come to Britain in 1944 to serve in the Air Force and frequently represented Britain in international competition. Randolph Turpin, born in Leamington in 1928 of a father from British Guiana, briefly became middleweight world boxing champion when he beat the American Sugar Ray Robinson in 1951. The Indian subcontinent preferred other sports. Half a century before, while studying at Cambridge, the flamboyant aristocrat Kumar Sri Ranjitsinji, Jam Sahib of Nawanagar, usually known as 'Prince' or 'Ranji', was the first non-white to represent England in international sport, going on to play for his adopted country at cricket 15 times, and his nephew Duleepsinji continued the tradition.

During Britain's post-imperial transformation, a whole new meritocracy emerged, inside and outside the sporting arena. In 1948, W. Arthur Lewis, born on the Caribbean island of St. Lucia in 1915, became Professor of Economics at Manchester. (Salam and Lewis met in Stockholm in 1979 when they both received Nobel prizes.) Britain was eager for skills and talent at all levels. Salam's area of science had no tradition to build on: modern field theory had only come into existence in the immediate post-war years, and needed agile young minds, unencumbered by pre-conceived ideas.

In 1954, Salam was thrown an unexpected personal lifeline. When Max Born retired from his Edinburgh chair and moved back to Germany, his replacement was Nicholas Kemmer, leaving vacant the prestigious post of Stokes Lecturer in Mathematics at Cambridge [12]. Salam had not applied for the post. Kemmer, guilty at having almost refused Salam as a research student a few years earlier, had pointed to his possible successor[13]. (This did not completely assuage Kemmer's conscience: in 1971, Salam's first honorary doctorate outside of his home country was awarded at Edinburgh, to Kemmer's 'private pleasure'[14].) While it was unusual, even remarkable, to make such an offer to a scientist from the Indian subcontinent, not that long before, British universities had benefited when a whole generation of posts had been taken up by intellectual refugees from Nazi Germany. Kemmer himself had been one.

The job offer brought a prestigious invitation to become a fellow of Trinity College, just down the road from St. John's, and with its glittering tradition, from Newton to Hardy. But Salam preferred the familiar surroundings of the college that had first sheltered him in 1946. With a British university lectureship and a Cambridge fellowship, Salam was no longer on the breadline. He would earn £450 annually as a university lecturer and £300 as a Fellow of St. John's College, together with £50 allowances. While he was happy to escape from the research isolation and religious persecution of his home country, he did not want to burn his bridges. Before leaving Lahore, he obtained three years leave of absence from his employers, the Government of the Punjab, during which time he would be able to return. This arrangement also covered a special payment of 180 rupees per month, which went to his family in Pakistan, for a member of staff 'on deputation' to Cambridge[15]. Although Salam had encountered problems in Lahore, others had realized that he was a phenomenon. In 1951, Mian Afzal Hussain, Chancellor of Punjab University, had written to Salam's father 'Men like Abdus Salam do not belong to any community or country. Their place is among the most brilliant in the world and therefore they belong to the entire humanity.... Wherever Abdus Salam has the facilities for his work he should stay there and Pakistan should help him.'[16]

In December 1953, Salam and his family moved to Cambridge, to live in an apartment on Bridge Street, near St. John's College. His father had just retired, and Salam sent money to Pakistan to augment his father's meagre government pension. Back in Britain, Salam was living in a country that he knew well by this time, and had mastered its language better than most of its inhabitants. This was not the case for his wife, Amtul Hafeez, who was having to cope with a strange country, a cold winter and a three-year old daughter. Salam himself encountered new challenges: one immediate obstacle was learning to drive a car, where he was not able to reverse the vehicle to the examiner's satisfaction, and failed his test at the first attempt. It startled him – the first time in his life that he had known failure. When not behind a steering wheel, he had to teach quantum mechanics and electricity and magnetism to undergraduates. With Dirac visiting the Institute for Advanced Study in Princeton that year, Salam had to instruct students in basic quantum mechanics. Having to deputize for the quantum master was an honour: Salam had himself attended Dirac's lectures just a few years previously. In his lectures, he refused to descend to a lowest common denominator level, preferring instead to keep his distance. In addition to giving lectures, Salam had to be available for student tutoring six hours per week. Here, the arcane mysteries of the Cambridge mathematics tripos course were still fresh in his mind. Peter Landshoff, later to become professor at Cambridge's new Centre for Mathematical Studies, recalled 'The first undergraduate supervision he gave us, coming straight from school, was very exciting. He would always begin by asking us what difficulties he wanted us to sort out, and then answer clearly and rapidly, so supervisions with him rarely lasted more than ten minutes, instead of the hour we paid for.' Salam had discovered that the heavy teaching load cut into his research time[17].

Salam's second daughter, Asifa, was born in November 1954, soon after Salam's younger brother Abdul Majid arrived at St. John's as an undergraduate, studying natural sciences. Natural sciences means some work in mathematics, but Abdul Majid was not formally taught at Cambridge by his brother. (After completing his degree, Abdul Majid returned to Pakistan as a manufacturing chemist, where he helped the country reinvent the manufacture of penicillin, at sixteen times the world market price. The episode showed Salam the inequities of the world technology market[18]. Later, Abdul Majid transferred to the United Nations International Development Organization, UNIDO, in Vienna as a technical advisor for the pharmaceutical industry. In Vienna, Salam met his brother regularly on visits to the headquarters of the International Atomic Energy Authority.)

At Cambridge, Salam now had to kick-start his research. To overcome his inertia after several years of isolation, he called on his friend Paul Matthews nearby in Birmingham. Together they had begun to recast the mathematical formalism of field theory during one of Salam's summer visits from Lahore. Now their objectives were more ambitious, consolidating the mathematical foundations that had been laid by Feynman and Schwinger. Proud of their sophisticated new mathematical scheme, which reproduced their earlier results more elegantly, they wrote their paper and asked Peierls, Matthews' boss at Birmingham, to write a covering letter for *Physical Review.* They were totally deflated when the paper was rejected, the referee alleging that the paper read more like 'a chapter of a book'. The referee who judges papers submitted for publication in a learned journal is supposed to be anonymous, but through a chink in the curtain the referee was discovered to be the authoritative Freeman Dyson. The paper was eventually published in the Italian journal *Il Nuovo Cimento* in July 1955. Salam was overjoyed when Feynman referred to it at a major physics meeting in 1955, and even more so when news of what had been achieved spread to the mathematical world. Salam was ecstatic when invited to present the results at a mathematical seminar at the Ecole Polytechnique in Paris, although some of the purists in the audience found the treatment sacrilegious[19].

This particular work, now largely forgotten, reflects Salam's pride in his mathematical ability and the pleasure he derived from difficult but elegant formalism. Scientific research is not a matter of sitting in a garden waiting for an apple to fall. As the US inventor Thomas Edison said ; 'Genius is one per cent inspiration and ninety-nine per cent perspiration'. The mathematics that physicists use is sometimes ugly and inexact, forced to fit into a framework for which it was not designed. Salam knew this and the mathematics of his physics could be contrived and makeshift. But he gained added pleasure when there were no such compromises. Matthews and Salam referred to their elegant reformulation of Feynman's methods as 'English' integral calculus, in contrast to Feynman's 'French' approach. To enliven the tedium of their calculations, Salam and Matthews would adopt ribald names for their mathematical quantities and equations. One of their joint efforts in field theory was privately dubbed *The Sphinx's back passage*, after a famous limerick.

Despite their best efforts, Salam and Matthews' theories were ineffectual and increasingly unfashionable. To take their place, a new technique – 'dispersion relations' – attempted to exploit the intrinsic mathematical properties of the quantum equations. If it were mathematical, then Salam could use it. Putting field theory to one side, he developed a more general approach to dispersion relations[20], presaging methods that for the next decade would usurp field theory as the premier research tool. Although field theory remained his research love, it is ironic that Salam, anxious to remain at the research forefront, helped contribute to its fall from grace in the mid-1950s. Although many soon forgot field theory, Salam did not, and carefully kept it within reach.

At St. John's College, Salam now regularly met Dirac, who he had previously admired from a distance. Every Tuesday, Dirac drove into College in his two-seater for dinner at High Table. 'Once Dirac asked me if I thought algebraically or geometrically. I did not know what he meant, ...' but after further questioning, Dirac said 'Precisely as, I thought. You think algebraically as most people in the Indian subcontinent do.' Dirac, it appeared, thought geometrically[21].

At Cambridge, Salam had inherited Kemmer's crew of research students, one of whom was Walter Gilbert. Born in 1932, Gilbert moved to Cambridge (England) after a first postgraduate year at Harvard, studying the theory of elementary particles. In 1955, Salam and Gilbert worked together on techniques to further extend the reach of Salam's dispersion relation approach. While working for Salam, Gilbert met James Watson, then working with Francis Crick at the Cavendish Laboratory. Crick and Watson's discovery of the double helix structure of DNA molecules using X-ray diffraction analysis in 1953 was one of the biggest scientific discoveries of the twentieth century. Although Walter Gilbert continued to work on theoretical physics, completing his PhD in 1957 and going on to work with Julian Schwinger at Harvard, his interest in molecular biology had been sparked. Linking up again with Watson, now also at Harvard, he joined an experiment trying to identify messenger RNA, a short-lived RNA copy of a DNA gene, which serves as a carrier of information from the genome to the ribosomes, the factories that make proteins. After his heroic discovery in 1966 of the protein 'repressor' molecule that controlled the action of genetic material, Gilbert went on to develop techniques for determining the molecular content of DNA, for which he shared the 1980 Nobel Prize for Chemistry. After founding a new company, Biogen, Gilbert became one of the main proponents of the Human Genome Project. Salam was naturally proud that one of his former students had gone on to attain Nobel status, and was not at all embarrassed that it was for achievement in a totally different field. For Salam, this was a valuable example of how physics could seed new developments in other fields, a theme that was to become one of his driving ambitions.

Another student inherited from Kemmer was Ronald Shaw, who did his first degree in mathematics at Cambridge while living in a room in Trinity College previously occupied by Freeman Dyson. Shaw had thought that his postgraduate research work would be supervised by Kemmer, who had at a similar stage in his career been supervised by Wolfgang Pauli. Shaw relates how Kemmer told him that, in his first week, Pauli had given him an extremely tough problem to investigate[22]. 'Kemmer was so dismayed that he very nearly gave up physics completely. So, to protect me from any similar dismay, Kemmer decided not to suggest any problems in my first year of research. Instead he guided me through the occasional Pauli paper, and made various suggestions (Schwinger, Feynman, Dyson, ...) of other papers to read. This suited me very well, as I liked to work on my own, following up my own ideas. Eventually, in early 1954, I became a research student of Abdus Salam. Salam's tendency was at the other extreme from Kemmer's: Salam was buzzing with research projects, often involving nuclear physics of which I was woefully ignorant. Consequently, I tried to keep away from Salam as much as possible, and to carry on following up my own ideas.'

Shaw submitted his PhD dissertation in September 1955. It consisted of two parts, each self-contained. The first was about the mathematics of special relativity and its possible implications for the types of elementary particles that could exist. The second part, called 'Some Contributions to the Theory of Elementary Particles' was easy to overlook, because its contents were only listed after the end of Part I, and itself was split into three parts, the third being called 'Invariance under general isotopic spin transformations'. A footnote in the dissertation reads: 'The work described in this chapter was completed, except for its extension in Section 3, in January 1954, but was not published. In October 1954, Yang and Mills adopted independently the same postulate and derived similar consequences.' The paper of C.N. Yang and Robert Mills was one of the most influential of modern theoretical physics. With it, field theory, which had been stuck in the doldrums for several years, was relaunched on a new course. Shaw continues 'But although their publication date was in 1954, Yang and Mills must have priority since it seems that their research was completed in 1953. My idea came to me in a flash while reading a manuscript of Schwinger's, which I found lying around in the Philosophical Library in Cambridge. I showed my generalization to Salam in early 1954, but in a rather disparaging way. Later in 1954, Salam showed me the paper by Yang and Mills. Salam still wanted me to publish my contribution, but I never did.' Subsequently Salam frequently referred to Shaw's work in the same breath as Yang and Mills, but realized how sceptical others were[23]. This oversight underlined Salam's own resolve to publish all his new proposals as an insurance policy. Unlike the highly principled Pauli, who went to great pains not to commit himself to ideas of which he was unsure, Salam decided it was better to publish somewhere and be wrong than not to publish at all and run the risk of losing credit.

In 1955, Salam's former Lahore student Riazuddin arrived at Cambridge to join the small but enthusiastic graduate student group. Adrift in the Cambridge throng, he appreciated the warm welcome at Salam's room in St. John's, where Salam would always find time for him, despite apparently having been absorbed in his work. Riazuddin also admired the way Salam could immediately resume his work after any interruption as though nothing had disturbed him.

Another of Salam's initial Cambridge research students was John Polkinghorne, who later wrote that Salam was 'a prolific generator of ideas. Salam has about him an air of uncontrolled intellectual fertility. Some of his ideas have been very good indeed.... but some of them have been...less inspired. People with this kind of gift and temperament.... function best when they have.... a strong and more cautious personality, able to...act as an intellectual filter and scientific conscience. Paul Matthews was for many years a partner of this kind.... This fecundity did not impinge on me very much when I was a research student. Salam mostly left me to pursue what interested me'.[24] Despite this, Salam and Polkinghorne collaborated in an interesting joint venture to look at the possible symmetry implications of subnuclear particles. Eventually becoming Professor of Mathematical Physics at Cambridge, Polkinghorne resigned this prestigious position in 1979 to pursue theological studies, becoming a priest in 1982. Since then, he has concentrated on bringing scientific rigour to Christianity, reconciling modern science with traditional interpretations of Christian scripture and dogma. In 1997 he was knighted for his distinguished services to science and religion, and in 2002 received the prestigious Templeton Prize, awarded 'to encourage and honour those who advance spiritual matters'.

Polkinghorne also said that 'Salam's exuberance extended to his lecturing style. At a conference, people would always be anxious to learn what he was thinking about. His innovative mind would be in gear until the last moment and he would turn up with a disreputable collection of hastily scribbled transparencies. One by one they would have been baffling enough, but Salam put the next one on before he had taken the previous one off. One got an impression of intellectual excitement, if not always a clear notion of exactly what the excitement was about'.[25] Some twenty years later, at a physics meeting in Britain, Salam's research student Christopher Isham (later to become Professor of Theoretical Physics at Imperial College) was deputizing for Salam, and was just about to embark on a talk illustrated by a huge pile of transparencies. At the last minute, Salam arrived and announced that he would take over. As Isham stepped aside, a worried Salam demanded 'Where are all the rest of my transparencies?'. Isham reached into his bag and handed over about twice as many as he had been going to use[26].

The clouds that had been hanging over the future of quantum field theory had not dispelled, but in the lull before the new Yang–Mills whirlwind, researchers had become used to the lack of quantitative sunlight. In this intermission, a new shape began to emerge on their radar screens, at first blurred and mysterious. In weekly coffee sessions in the Christ's College rooms of lecturer James Hamilton, Cambridge theoretical physicists speculated on the possibility that certain subnuclear reactions called for a reappraisal of basic principles. What had been thought sacrosanct perhaps was not, thereby answering the rhetorical question Peierls had put to Salam – why a massless neutrino?

REFERENCES

1. Matthews, P.T., Salam, A., *The renormalization of meson theories*, Rev. Mod. Phys., 23, 311–4 (1951)
2. Singh, J., *Abdus Salam, a biography* (New Delhi, Penguin, 1992)
3. Ghani A., *Abdus Salam, a Nobel laureate from a Muslim country* (Karachi, published privately, printed Ma'aref, 1982)
4. Matthews, P.T., Salam, A., *Renormalization,* Physical Review, 94, 185–91 (1954)
5. Enz, C., *No time to be brief – a scientific biography of Wolfgang Pauli*, (Oxford, OUP, 2002)
6. Lai, C. H., Hassan, Z. (ed.), *Ideals and realities, selected essays of Abdus Salam*, (1984, Singapore, World Scientific), 211
7. Salam, A., The Field Theory of Superconductivity, Prog. Theor. Phys., 9, 550–4 (1954)
8. Pais, A., *Einstein lived here*, (Oxford, OUP, 1994) 117
9. Ziman, J. *Delight in Abdus Salam's company*, in Hamende, A. M. (ed.), *Tribute to Abdus Salam*, (Trieste, ICTP, 1999)
10. Vauthier, J. *Abdus Salam, un physicien* (Paris, Beauchesne, 1990)

11. Lai, C. H., Hassan, Z (ed.), *Ideals and realities, selected essays of Abdus Salam*, (1984, Singapore, World Scientific) 89
12. Named after the Irish mathematician George Gabriel Stokes (1819–1903), Lucasian Professor of Mathematics at Cambridge, and famous for Stokes' Theorem and the Navier–Stokes equation. Previous holders of the post included James Jeans and Max Born.
13. ICTP archives A28
14. N. Kemmer, 'The Cambridge Days', in *Abdus Salam as we know him*, ed. S.M.W. Ahmad (Trieste, ICTP, 1990)
15. Ghani (1982)
16. Abdul Hamid (brother), English synopsis, dated September 1997, of Urdu biography
17. Vauthier (1990)
18. Lai, C. H., Hassan, Z (ed.), *Ideals and realities, selected essays of Abdus Salam*, (1984, Singapore, World Scientific) 155
19. Salam, A., *A life of physics*, in Cerderia, H. A., Lundqvist, S. O., (ed.) *Frontiers of physics, high technology and mathematics* (Singapore, World Scientific, 1990)
20. Salam A., *On generalised dispersion relations*, Nuovo Cimento 3 424–9 (1955)
21. Salam, A., *The Unification of fundamental forces* 1988 Dirac Lecture, (Cambridge, CUP, 1990)
22. http://www.hull.ac.uk/php/masrs/reminiscences.html
23. For other references to Ronald Shaw's work, see:
    O'Raifeartaigh, L., *The dawning of gauge theory* (Princeton, N.J., Princeton University Press, 1997)
    Taylor, J. C. (ed.), *Gauge theories in the twentieth century*
    (London, Imperial College Press 2001)
24. Polkinghorne, J., *Beyond science – the wider human context*, (Cambridge, CUP, 1996)
25. Polkinghorne, J. (1996)
26. E-mail K. J. Barnes

8

'Think of something better'

Salam received his Nobel Prize in 1979 for 'contributions to the theory of the unified weak and electromagnetic interactions between elementary particles'. About a hundred years before, James Clerk Maxwell could have earned a Nobel prize for 'contributions to the theory of the unified electric and magnetic interactions, now called electromagnetism', but there were no Nobel prizes before 1901. The force of electromagnetism is a subtle one. It can drive mighty machines: but the electricity that drives these machines is harvested from the force that holds tiny atoms together. Salam showed how electromagnetism is linked to another subatomic force, the 'weak nuclear interaction', much more subtle than electromagnetism. It acts deep inside the nucleus at the heart of the atom, yet it provides vital sparks without which the Sun's thermonuclear furnace would not ignite. When Salam first learned about electromagnetism at school in Jhang, there was no electricity to be had there: the teacher told his pupils that to encounter it, they should take the train to Lahore, several hundred miles away. The Sun beat down on Jhang, but nobody knew that its fierce heat was indirectly the result of the weak interaction. To learn about this most subtle of Nature's agents, Salam had to embark on a much longer journey.

After the theory of quantum electrodynamics had been put together in the late 1940s, the next objective had been to use this theory as a template for a more ambitious one that explained the forces that hold the nucleus together (unimaginatively called by physicists the 'strong force'). This bid had initially failed, thwarted by the strength of the nuclear force, which rendered calculations impossible, and by the unexpected discovery of so many unexpected subnuclear particles. Instead, it would be the shadowy weak force that would provide the setting for the next breakthrough.

Thunder and lightning are always impressive. Even the words themselves seem to roll and flash across the page. The Holy Qur'an, (Sura 13, verses 12 and 13) say: 'He it is, who shows you the lightning to induce fear ... and the thunder glorifies Him ....' Lightning is what happens when accumulated atmospheric electricity can no longer be contained and overflows to the Earth below, wrenching apart the atoms of the atmosphere on its way. The deafening crash of thunder is the displaced air rushing back to its original position. Lightning is a giant spark. Normally gases like air do not conduct electricity, but when the voltage is pushed high enough, as in a thunderstorm, the atoms of the gas break apart and a current can pass.

Nineteenth-century scientists found other kinds of sparks when electricity was passed through air or some other gas in a glass tube. Incandescent lighting, produced by passing electricity through a suitable filament–gas arrangement, had only been discovered in 1878, so anything that produced light, or any kind of rays, was sensational. When the air in the tube was rarified, all that remained of the spark was a dull glow. If objects were sealed into the tube, the glow of mysterious 'cathode rays' could cast shadows, generate heat or even turn a tiny wheel. Their mystery evaporated in 1897 when J. J. Thomson at the Cavendish Laboratory in Cambridge showed that they were due to residual atoms of gas snapping apart, releasing tiny particles of negative electricity – electrons.

But even before the cathode ray mystery had thus been solved, other kinds of atomic radiation had turned up – Germany had its 'Röntgen rays' (X-rays) and France had its 'Becquerel rays'. To add to the confusion, Pierre and Marie Curie in Paris found still more, which they named 'radioactivity'. In Cambridge, the New Zealand research student Ernest Rutherford calmly analysed this disarray, and found two distinct kinds of radioactivity: alpha radiation, which could be easily blocked by a sheet of paper; and beta radiation, a hundred times more penetrating[1]. He subsequently focused on alpha radiation, discovering it was made up of helium atoms that had been stripped of their electrons. These 'alpha particles' became Rutherford's stock-in-trade. Others found that beta rays carry negative electric charge and behaved in the same way as cathode rays and Thomson's electrons. Beta rays, or more precisely, beta particles, were electrons.

As the nineteenth century merged into the twentieth, civilization appeared to wake from a torpor. Even a hundred years later, the impact of these sudden innovations still looks impressive. In the biggest advance in transport technology since the invention of the wheel, vehicles driven by internal combustion engines replaced horse power.

The telegraph and then the telephone brought instantaneous messaging: Marconi's radio waves would soon bring more. 'Modernism' became the collective envelope for what Picasso and Braque did for art, Stravinsky in music, Le Corbusier in architecture, Kafka, Proust and Joyce in literature, Brecht and Ionescu in theatre. Physics was modernistic too, with electrons and radioactivity, and soon Max Planck's quanta and Albert Einstein's relativity.

With so much happening, scientists were confused. When the time comes to write textbooks, developments can be strung together in a logical order. But scientific research does not open like a textbook. Its pages are not numbered, and there is no guarantee that discoveries will happen in the right sequence. Scientists have to struggle with prejudices and preconceptions, just as Columbus did when he landed in America and thought he was in the 'West Indies'. At the beginning of the twentieth century, physicists knew only that cathode rays (electrons), X-rays and radioactivity were all due to things that happened inside the atom.

Under electric stress, atoms fray and their electrons leak out as cathode rays. After all these negatively charged electrons have been stripped off, what remains has to carry net positive electric charge. This was where radioactivity came from. At first, physicists imagined the atom as an electrical cake, with negatively charged electron fruit evenly dispersed in some all-enveloping positively charged sponge. In 1911 Rutherford's alpha particle experiments revealed otherwise. There was no atomic cake. Instead, the atom was mostly empty – 'a miniature solar system' is the traditional description – with minuscule electron planets orbiting a heavy, compact nucleus. This tiny atomic kernel was the source of radioactivity. Just as they had learned about atoms by shining atomic light through prisms and studying the emergent colours, scientists eager to learn about nuclei now turned to nuclear spectra.

They saw alpha particles with distinct energies, reminiscent of the sharp spectral lines of atoms. Inside the nucleus, protons appeared to sit on the rungs of a ladder-like framework, as electrons did inside the atom. However, the beta electrons were more enigmatic. Electrons had appeared to live in the atomic suburbs and initially had been excused any nuclear role. Now physicists saw that electrons also emerged from the nucleus. Why? And this was not the end of the puzzle. If the electrons came from inside the nucleus, their energies should be stepped, reflecting the energy structure of the nucleus itself, just like the alpha particles. Instead, beta electrons had a smooth range of energies. It looked like beta electrons slid down slopes, rather than climbing down ladders. How could the atomic nucleus have an energy ladder for some particles but not for others?

The quantum revolution had already introduced disconcerting ideas like the Uncertainty Principle. A single atom sits quietly minding its own business in the dark, until a curious investigator wants to pinpoint exactly where the atom is. The investigator cannot see in the dark, so must illuminate the atom in some way. But the mere act of switching on a light creates photons. These bump into the tiny target atom, which immediately recoils away. Observation is a self-destruct mechanism – attempting to observe something actually messes up the observation! Just as they began to learn how unfamiliar quantum mechanics could handle these enigmas, physicists wondered whether even worse things happened inside the nucleus.

Scientists both revered and feared Wolfgang Pauli. Born in Vienna in 1900, he matured with the baffling ideas of modern relativity and quantum theory. Whenever anything radically new appeared, Pauli seemed immediately to understand it more profoundly than anyone else. He saw that the electron is not simply a dot of negative electricity, but behaves instead like a tiny arrow that twists around its length, and the electron's behaviour depends on the direction of this spin. On the atom's energy ladder, only two electrons – one spinning clockwise, the other anticlockwise – can stand on one energy rung at any time. This was the Pauli Exclusion Principle, which dictated how many electrons can be fitted into each kind of atom. It was the ultimate explanation of why the natural chemical elements can be slotted into the grid of the Periodic Table.

Pauli made scientific rigour into a weapon, a touchstone both for his own work, and for his opinion of others, and was highly intolerant of the muddled thinking that often passes for contemporary research. His rigour was only matched by his mastery of irony: a criticism from Pauli could be the ultimate put-down. A candidate theory was so wide of the mark as to be 'not even wrong[2]'. To Werner Heisenberg, the discoverer of the Uncertainty Principle, Pauli once said 'It's much easier to find one's way if one isn't familiar,' but adding impishly 'But lack of knowledge is no guarantee of success.'[3] Pauli would never permit half-baked ideas, and exposing new proposals to him was seen by others as the ultimate testbed. The risk was that fragile egos could be punctured in the process. Salam too was to become a victim of Pauli's intellectual wrath.

Having introduced the revolutionary ideas of electron spin and the Exclusion Principle, Pauli's imagination seemed to have been exhausted and he retired into a hard carapace of orthodoxy. After the tumult of the early twentieth century, he felt that physics had been put on a firm foundation. The challenge now was to understand everything in terms of what was already known. But after making his greatest contributions to science, Pauli's personal life suddenly seemed to fall apart. His mother committed suicide, his wife ran off with another scientist, and he began drinking. 'It is easier for me to achieve academic successes than successes with women,' he complained[4], and spent many hours on the couch of pioneer psychoanalyst Carl Jung.

Pauli had thought of a way out of the beta electron energy dilemma, but it was too unconventional for him, and he pushed it into a dark corner of his mind. Invited to speak at a meeting on radioactivity in Tübingen in December 1930, the stressed-out Pauli preferred to stay in Zurich, go to a pre-Christmas student ball and drink wine. Perhaps he would find a new partner. To excuse himself to the physicists at Tübingen, he sent a letter, 'Dear radioactive ladies and gentlemen'...., in which he surmised that perhaps each beta electron fell off its energy rung on the nuclear ladder not on its own, but accompanied by another particle, so delicate as to be invisible, but which could nevertheless steal energy from the electron. 'I dare not in the meantime trust myself to publish anything about this idea', wrote Pauli, 'and address myself confidentially to you.' Pauli had let himself off the hook. In his ultraconventional mind, such an idea was not a valid scientific suggestion, just a pre-Christmas joke, a subnuclear particle that was invisible but could still somehow do things. He knew that if he heard such an outlandish idea uttered in public by another scientist, he himself would immediately condemn it as 'Quatsch' – worthless rubbish. Towards the end of his life, he said of the then hypothetical particle 'that foolish child of the crisis of my life – and which later itself behaved foolishly'.[5]

Pauli called his invisible particles 'neutrons', not knowing that the name had meanwhile been adopted for something else. Rutherford had long suspected that nuclei, as well as containing protons, also contain electrically neutral particles that live happily alongside the positively charged protons, like husband-and-wife pairs. In 1932, James Chadwick at Rutherford's Cavendish Laboratory discovered that Rutherford's neutrons were real. If Pauli's 'foolish' idea was to be useful, it first had to have a name. It was to be an Italian one. The prescient Orso Corbino in Italy had seen the momentous science going on in Britain, France and Germany in the early twentieth century, and decided it was time for his country too to make its mark. The man he chose to do it was Enrico Fermi, born in Rome in 1901, perhaps the greatest scientist Italy had seen since Galileo, three hundred years before. In 1933 Fermi found a way to describe the vital role that Pauli's 'foolish' particle played in beta decay. But with Pauli's option on the 'neutron' title now lapsed, Fermi christened it with the Italian diminutive 'neutrino' – 'little neutron'.

In 1938, Fermi was awarded the Nobel Physics Prize for his work on neutrons, rather than neutrinos, but only bought a one-way ticket to Stockholm. With a Jewish wife, and alarmed by increasing anti-Semitism, he emigrated to the United States. There, as the clouds of war gathered, he and many other nuclear physicists were marshalled together and told to develop an atomic bomb. Beta decay and neutrinos were temporarily forgotten. A vast new laboratory complex was built at Los Alamos in the New Mexico desert. Progress was swift: the first bomb was exploded at the desolate Jornada del Muerto, the first 'Ground Zero', two hundred miles south of Los Alamos, at 05.30 on 16 July, 1945. Fermi, watching the explosion several kilometres away, let a few pieces of paper fall from his hand. Floating away in the blast, they landed 2.5 metres from his feet, from which he calculated that the blast was equivalent to ten thousand tons of TNT.

But this was only the start, not the end of nuclear business at Los Alamos. The end of one war brought an uneasy peace that soon dissolved into a new 'Cold War', inviting new weapons of mass destruction. The first thermonuclear device was exploded at Eniwetok in the Marshall Islands on 1 November 1952. By then, physicist Fred Reines, working at Los Alamos, had had enough of bombs and wanted a less sinister way of making a living. One by-product of the new bombs, realized Reines, was Pauli's invisible neutrino. Neutrinos had been plentiful at Hiroshima and Nagasaki, but nobody noticed. After the war, Fermi had left Los Alamos to become physics professor at Chicago, but still visited the laboratory from time to time. There, in 1951, Reines met him, and their conversation went something like this[6]:

Reines: I have been thinking about detecting neutrinos, and the bomb may be the best source.

Fermi: (after some thought) Yes, that appears so.
Reines: It would need a detector with a sensitive mass of about a ton.
Fermi: That's right.
Reines: I have no idea how to build such a detector.
Fermi: Nor have I.

And that was the end of their conversation.

Designing a detector sensitive enough to register the arrival of flimsy particles but robust enough to withstand a nuclear explosion was difficult. After a few months, Reines, now working with Clyde Cowan, realized that fission reactors should also produce lots of Pauli neutrinos, and would be somewhat more user-friendly. In 1953, Reines and Cowan went to the reactor at Hanford, in the state of Washington, where a reactor complex had been built to supply plutonium for fission bombs. The pair assembled a bath-sized 300-litre tank of sensitive scintillating liquid, which picked up tiny flashes as particles passed through. Maybe a few of these would be due to neutrinos. In the clutter of flashes, Reines and Cowan saw what could have been a neutrino signal, but could not be sure. So they built a bigger detector, this time containing several tons of scintillator, and took it to a more powerful reactor, at Savannah River, South Carolina, which had been built to provide nuclear materials for hydrogen bombs, and watched it for about a year. By June 1956, the pair were convinced they had seen a neutrino signal, and sent a telegram to Pauli 'We are happy to inform you that we have definitely detected neutrinos . . .'. Unfortunately, the telegram was addressed to Zurich University, rather than the Swiss Federal Institute where Pauli worked, and anyway he was attending a meeting at CERN, the European particle physics centre in Geneva, where the announcement was finally read out[7]. Pauli, relieved that his 'foolish' prediction was no longer a nightmare, wrote to Reines and Cowan: 'Thanks for the message. Everything comes to him who knows how to wait'. Ironically the reply was overlooked in the general celebration, and didn't arrive until much later.

By the 1950s, beta decay had been classified as one example of a much wider class of ' weak nuclear interactions', so styled to contrast them with the 'strong nuclear interactions' that gripped nuclear protons and neutrons tightly together. These weak interactions were no longer confined to unstable atomic nuclei. In the 1930s, the menu of known subatomic particles was very short – protons, neutrons, electrons, photons, and perhaps Pauli's foolish neutrino. Twenty years later, the Greek alphabet had been mobilized to name new subatomic species found in cosmic rays and in new experiments at the huge particle accelerator laboratories. Many of these particles labelled by Greek letters were broken apart by the weak force, providing a huge new window on beta decay. But two different Greek-lettered particles, clearly differentiated by their subnuclear offspring, otherwise looked the same. It was a scientific Jekyll and Hyde – as though the same person, living at the same address, had two completely different weak interaction personalities.

In 1956, the dilemma was seized by the two young Chinese scientists, Tsung Dao (T. D.) Lee and Chen Ning ('Frank') Yang, who earlier had learned physics from Subramanyan Chandrasekhar in Chicago. To explain the particle schizophrenia, Lee and Yang wondered if quantum behaviour, celebratedly bizarre, could also be sensitive to direction, and could somehow differentiate between left and right. Pauli did not think so. For him, and for most other physicists, left and right was merely a matter of convention. Anything that can be built using mathematical screws that tighten clockwise could also be built using non-conformist screws that tighten the other way. What happens in the interior of the nucleus is left–right symmetric, thundered Pauli.

Objects that appear left–right symmetric, like people, can nevertheless operate in a right-handed or left-handed manner. Suppose, ventured Lee and Yang, that particles too could be in some way right-handed or left-handed. Imagine electrons filing through a turnstile that turns clockwise: those coming through will have a momentum to their right and emerge facing in that direction. For a turnstile rotating anticlockwise, the electrons will point toward their left. Looking hard at the way some particles decayed, there were tantalizing examples that this could be the case. But more definite proof was needed. A team at Columbia University, New York City, prepared a decisive experiment.

While this apparatus was still being built, Frank Yang described these new ideas in a talk on 'New Particles' at an International Conference on Theoretical Physics in Seattle from 17–21 September 1956. In the audience was Abdus Salam, 30 years old, and excited. After a couple of years as a lecturer at Cambridge, he was now getting job offers – a readership, a couple of rungs up the promotional ladder, at Liverpool; and a visiting professorship at Rochester, New York. But he knew that a more tempting offer would soon be coming. The new ideas of weak interactions possibly being sensitive to direction appealed to him, and he knew that

Reines and Cowan had just found the elusive neutrino. He had a lot to think about on the flight home from Seattle.

In those post-war years, the US Air Force had spare cash for pure research projects that it distributed to a few chosen university departments outside the US. This enabled university staff and students to use the Military Air Transport Service (MATS). These flights were primarily meant for US servicemen and their families transferring to and from distant assignments or going on leave. Overnight transatlantic flights in the pre-jet era were arduous, but much faster than the other cheap option, a transatlantic boat trip. Salam's MATS flight back to the UK in September 1956 was full of young children crying.

Despite being used to cramped conditions from his childhood, he could not sleep, and kept thinking about what he had heard in Seattle about weak interactions, especially Yang's suggestion that they could be sensitive to direction. The hallmark of weak interactions was Pauli's neutrino, now no longer hypothetical – a weak interaction is either caused by or produces one. Salam recalled the culmination of his Cambridge doctorate examination four years earlier. The ultimate arbiter of a doctorate thesis is an objective examiner who can ask the candidate penetrating questions about the thesis, or anything else for that matter. Salam's examiner had been Rudolf Peierls from Birmingham, who, according to Salam, had asked him in the examination why the mass of the neutrino should be zero. At that time, nobody knew whether neutrinos existed, let alone what their mass was. The only indicator was that they had to be extremely light. Salam's answer to Peierls question was not recorded, but he saw the irony of being asked a question to which the examiner himself most likely did not know the answer. 'During that comfortless night, the answer came,' related Salam later[8]. If the neutrino's mass is zero, Salam saw that Pauli's particle acts as a tiny corkscrew, drilling through space at the speed of light. Viewed from behind, a particle appears to spin one way. But if the particle is overtaken, it appears to move backwards and spin the other way. However, travelling at the speed of light, the neutrino cannot be overtaken, and always spins the same way. If the neutrino has no mass, its only label is its spin direction, so the reflection of a neutrino in a mirror is no longer a neutrino. Left–right symmetry is automatically shattered by massless neutrinos.

It was a Eureka moment. But to work out its consequences, Salam needed a few formulae that he did not carry in his head. After a sleepless night, he charged off the MATS flight, rushed to his office in Cambridge and worked out a few things. He saw that the massless neutrino would give big effects. Grabbing his calculations, he ran to the railway station and leapt onto a train to Birmingham to tell Peierls that he finally had the answer. One can imagine Peierls finding the excited young Pakistani unannounced on his doorstep. 'I do not believe left–right symmetry is broken in weak interactions at all,' he retorted, echoing Pauli. 'Thus rebuffed in Birmingham, like Zuleika Dobson, I wondered where to go next,' continued Salam, referring to the novel by Max Beerbohm, and displaying more of his diligent reading of English literature in Lahore. 'The obvious place was CERN in Geneva, with Pauli – the father of the neutrino – nearby in Zurich.'

It was to be one of Salam's big mistakes. He had been summoned by Pauli before, and had rushed to Bombay in 1952, but then he had not spoken about neutrinos. Why also was CERN the 'obvious' place to go? The Geneva laboratory had been established in 1954 by a consortium of Western European nations to build giant subatomic accelerators. These would rival those being constructed in the USA, and help wrest back for Europe the initiative in basic physics, which had shifted across the Atlantic during the Second World War. In 1956 the CERN laboratory was still under construction, and staff lived in barrack-like huts. But it had already established itself as a scientific transit lounge, where researchers could pause *en route* to whatever was their final destination, and talk. After Peierls had slammed the door on him, maybe someone at CERN would listen to Salam. He wanted to stake his claim and publish his idea in the scientific literature, but first wanted to air it in authoritative company. In 1949 Salam had immediately taken Matthews' renormalization problem to Dyson, and followed him from Birmingham to London. Salam, like a schoolboy eager to please his teacher, now wanted to show his massless neutrino idea to Pauli. The patient Dyson had been very helpful. Maybe Pauli would too.

But Salam did not know that Pauli had met the idea of massless particles some thirty years before. Dirac's equation for the electron, now regarded as a scientific monument, had initially been an embarrassment, introducing new components of the electron that did not correspond to anything that was known at the time. Hermann Weyl had introduced the idea of massless particles in 1929 to try to clear up some of the mystery surrounding the extra electron components before Dirac had dared to say that his equation predicted the existence of antimatter, in the form of antielectrons[9]. Pauli had been shocked to realize that Weyl's massless particles would break left–right reflection symmetry. For him, all such symmetries were sacrosanct, and immediately condemned Weyl's idea.

Pauli had heard the news of the neutrino discovery at CERN in that summer of 1956. Now he was at home in Zurich. Geneva and Zurich are not 'nearby'. Perhaps, once in Geneva, Salam would pick up the courage to get on a train to Zurich and knock on Pauli's door, repeating the experiment he had made on Peierls. But while Peierls was polite and paternal, the caustic Pauli was scary and inaccessible. Arriving at CERN, Salam met Felix Villars, who worked with Pauli and was just leaving for Zurich. Salam gave him a copy of his paper with the massless neutrino suggestion. Villars took it with him and immediately passed back the message from the oracle – 'Give my regards to my friend Salam and tell him to think of something better.' Quashed, Salam shelved his massless neutrino suggestion for two months. It was a decision he would later regret.

Pauli was by now the universally acknowledged Chief Justice of Physics. He did not approve of Salam's idea because of his long-standing conviction that physics could not possibly be sensitive to direction. He knew that an experiment was testing the hypothesis, but did not see the point. Towards the end of 1956, several months after seeing Salam's idea, Pauli was still adamant: 'I do not believe that the Lord is a weak left-hander, and am ready to bet a very large sum that the experiments will give symmetric results'[10].

The team at Columbia University were now carefully monitoring beta decay electrons. Radioactive nuclei acted as tiny magnets, each one a nuclear 'compass', sensitive to the direction of the external magnetic field. The experiment was run at cryogenic temperatures to calm thermal quivering. Just a few days after Pauli had stated that he was prepared to wager that nothing unexpected would be seen, the team was amazed to see that most of the beta electrons came out on one side. When the direction of the magnetic field was changed, the beta electrons emerged on the other side. It was a big effect from a simple experiment that could have been done at any time since beta electrons had first been seen half a century before, but nobody had bothered. The weak interaction is left–right sensitive. In physics-speak, weak interactions violate 'parity'. In the afternoon of 15 January 1957, the Physics Department of Columbia University called a press conference, and the next day the New York Times ran a headline 'Basic Concepts In Physics Reported Upset In Tests'. Even before, Salam could have heard about these developments from John Ward, then at the University of Maryland, who wrote to Salam to say that Einstein 'must be spinning in his grave. Clockwise presumably'[11].

Pauli was stupefied – On 27 January he wrote 'Now the first shock is over and I begin to collect myself again ... On Monday 21st the mail brought me three experimental papers [about the discovery of parity violation] ... the same morning I received two theoretical papers .... The latter was essential identical with the paper by Salam which I already received six to eight weeks ago..... It is good we did not make a bet. It would have resulted in a heavy loss of money (which I cannot afford). I did make a fool of myself (which I can afford) – incidentally only in letters or orally and not in anything that was printed. But now the others have the right to laugh at me.'[12] Quite apart from the startling implications of the revelation that something so fundamental had been overlooked for so long, Pauli was worried that his reputation as the oracle of physics had been shattered.

Pauli, who had mercilessly criticized generations of young physicists, was also embarrassed that he had discouraged Salam, who had come a long way to see him in Bombay, and then who had come to him with an idea that was right but that Pauli had thrown back at him. Pauli was the chairman when Salam spoke about the new developments in a meeting at the UK Rutherford Laboratory in December 1956. At the end of the talk, Pauli apologized to Salam[13], who eventually published his massless neutrino idea in the January 1957 edition of the Italian physics journal *Il Nuovo Cimento*. The manuscript had been received on 15 November 1956, several months after the idea had been rejected by Pauli. Salam should have sent it immediately to the prestigious *Physical Review*, run by the American Physical Society, where Salam had published his 1949 renormalization breakthrough. There was even a special outlet for fast-breaking developments – *Physical Review Letters*. But after being discouraged by Pauli, Salam agonized for two months before deciding to hedge his bet.

In the US, Lee and Yang, who had continually been aware of the status of the crucial test experiment, had come to the same conclusion as Salam about the role of the massless neutrino in beta decay. Their proposal was published in *Physical Review* in March 1957. In the Soviet Union, the influential physicist Lev Landau had also come up with the idea and published it, in English, at the same time. At the end of his paper, Salam said 'The author is deeply indebted to Professor Peierls who first suggested investigating the consequences of the requirement [that the mass of the neutrino is zero]'. This was written in 1956. In his 1979 Nobel lecture, his elaborate tale of taking the train to Birmingham accused Peierls of not welcoming at all his suggestion. The two versions did not tally. This piqued Peierls, who wrote to Salam, recalling other discussions that did not corroborate the Nobel Lecture anecdote[14]. According to Peierls, Salam had already proposed to him the idea of a joint paper on the implications of massless neutrinos, even before the discovery of parity violation. Peierls had refused. Like Pauli, he did not wish to even contemplate the implication that left–right symmetry is violated. Salam's Nobel lecture tale that the role of the massless neutrino had come to him in an inspired flash in an airplane above the Atlantic was a romantic exaggeration, but he knew that he had to give credit to Peierls.

The world of physics now had to reconcile itself to the new awareness of left–right sensitivity. On 24 January 1957 a repentant Pauli wrote to Salam – 'Our old friend parity died'. In his 1979 Nobel lecture, Salam says 'Thinking that Pauli's spirit should now be suitably crushed, I sent him two short notes I had written in the meantime'[15]. These extended the implications of zero-mass neutrinos for weak interactions. 'Pauli's reaction was swift and terrible,' recounted Salam, who goes on to quote Pauli: 'I am very much startled on the title of your paper "Universal Fermi Interaction". For quite a while I have made for myself the rule, if a theoretician says *universal* it just means pure nonsense....And now you, my dear Brutus, come with this word..... I have not seen your paper, but I have some small hope......that you have already withdrawn it.' In fact Salam had not done so, but after attending a meeting at Rochester, New York, in the summer of 1957 had come himself to the conclusion that it was deficient. On his way back to New York City, Salam dropped in at the editorial offices of the *Physical Review* and snatched back the paper, which had already been marked up for typesetting. If it had been published, the paper would have been recognized as a step towards the acknowledged version of the modern 'V-A' theory, published in 1958. Salam later regretted snatching back the paper and losing credit for another research milestone[16].

During 1957, Pauli frequently wrote to Salam, pointing to unpublished new results that had been leaked to him. People still liked to use

Pauli as a sounding board. Salam's replies were now frenzied. These airmail exchanges were as hastily written as the e-mails of today, with Salam in particular waking up the next morning and realizing that he had made a mistake that was already airborne on its way to the oracle in Zurich. Two such hasty letters in early 1957 were followed up by even hastier apologies/corrections. For his part, Pauli had no worries about asking Salam to reserve him a hotel room in London or a sleeper train berth from London to Edinburgh[17].

The new awareness of the left–right sensitivity of Nature prompted Salam to ask a classicist colleague if ancient literature had any portents; 'I asked him if any classical writer had ever considered giants with only the left eye. He confessed that one-eyed giants had been described, and supplied me with a list of them; but they always [like Cyclops in Homer's *Odyssey*] sport their solitary eye in the middle of their forehead. What we have found is that space is a weak left-eyed giant'[18].

In 1957, Lee and Yang were awarded the Nobel Physics Prize for 'their penetrating investigation of the so-called parity laws, which has led to important discoveries regarding the elementary particles.' In an area where the interval between the discovery and the award can be anything up to forty years (in 1995 Reines was so honoured for his 1956 neutrino discovery), Lee and Yang's achievement was startling by its promptness. The Nobel had been awarded to Lee and Yang for setting the scene for parity violation in weak interactions. Salam had entered only in Act II, realizing why parity should be violated before any effect had been seen, and that it would be a big effect. He had not waited for the proof from experimental discovery to emerge. Had he not been muzzled by Pauli, he would have published months earlier. Some gossip and garbled reports of the Nobel award even mentioned Salam's name. Friends sent Salam their congratulations and looked forward to a share of a Nobel prize. The Prize for physics can be awarded to up to three people. By now many people were on the scene, but clearly Lee and Yang had made the first move.

Salam resolved not to make the same mistake again. There was no harm in submitting revolutionary new ideas for publication: if they were not right, they could still influence others. It was up to the journal's editor to decide whether to publish it or not. Salam had a lot in common with Lee and Yang. All three were post-war intellectual immigrants from Asia: Salam from India to Britain; Lee and Yang from China to the US. All had passed through Princeton's Institute for

Advanced Studies. After Lee and Yang's achievement, Salam realized that the Nobel Prize for physics was within his reach. 1957 was one of the few times that it had gone to researchers who had not been born in Europe or the United States[19]. Asians could go to Stockholm. Salam had almost got to the prize with the massless neutrino: all he had to do now was to follow Pauli's advice – 'think of something better'.

While all this was going on, Salam was also preparing to move to a new job as Professor of Applied Mathematics at London's Imperial College of Science and Technology. Imperial and Cambridge could not have been more different. The University of Cambridge, one of the oldest universities in the world, has a history going back to 1284. Imperial College had only been established in 1907 as a new school of the University of London, incorporating the Royal College of Science, the City and Guilds College (mainly engineering) and the Royal School of Mines (geology and mineralogy). These three colleges were scattered on a pleasant campus area in South Kensington, just south of Hyde Park, which had been developed in the second half of the nineteenth century using the proceeds of the Great London Exhibition of 1851. It attracted such intellects as T. H. Huxley. H. G. Wells, the father of science fiction, had been his student. As well as institutes of higher education, the campus also included several major museums and cultural centres. The establishment of Imperial College at the beginning of the twentieth century had been a response to the growing awareness that heavily industrialized countries required a supply of trained scientists and technologists. To assure its future, Britain needed such specialists as much as it needed battleships. Similar universities were being established in the USA and in Germany. In 1932, G. P. (George) Thomson, the son of J. J. Thomson, the discover of the electron, became Professor of Physics at Imperial, and five years later shared the Nobel Prize for Physics for the discovery of electron diffraction, bringing new prestige and placing Imperial at the forefront of science.

After the Second World War, dragging Britain out of the mire of that conflict needed a greater investment in science and technology, and in the early 1950s Imperial College had been identified as the spearhead of a major national effort, the plan being to double the size of the college in ten years. When Thompson retired, his place was taken in 1953 by Patrick Blackett, who had won the 1948 Nobel Physics Prize for his work in imaging and recording subnuclear reactions, a career that had begun with Rutherford in the Cavendish laboratory in Cambridge in the 1920s. (The announcement of this award coincided with Salam's decision as a student at Cambridge to switch from mathematics to physics.) During the war, Blackett had served on the UK Air Defence Committee. With Britain's post-war socialist government, Blackett, a militant socialist and a supreme technocrat, had become very influential, both in Britain and overseas. In 1947 he was invited by Indian Prime Minister Jawaharlal Nehru to advise on the research and development needs of the Indian armed forces, becoming a frequent visitor to the country, and a close associate of Homi Bhabha.

Blackett understood well the objectives of Imperial's physics department, which would soon move into a new purpose-built block (eventually to become the Blackett Laboratory). Larger premises meant more staff, and Blackett set about headhunting. Although himself an experimental physicist, he understood well the synergy between experiment and theory in modern science. Blackett was also well travelled and mingled with influential scientists from all over the world. By this time Salam had carved out both his own personal reputation and a reputation for a dynamic research school. The work of his research students at Cambridge had already been noticed. Blackett was no longer active in subnuclear physics, but knew well those in the United States that were, like Hans Bethe, J. Robert Oppenheimer and Victor Weisskopf, who had all been impressed by Salam during his short visit to Princeton. Perhaps anxious to do something for the other new country in the subcontinent, Blackett, prompted by Bethe, offered Salam the job of Imperial's Professor of Applied Mathematics with a salary of £3000 per year. Salam relates how Blackett, a tall, imposing figure, came to see him in his room at Cambridge. Opening the door, he looked directly at Salam and demanded 'Do you want a chair?' When Salam replied affirmatively, Blackett said 'It is done.' However, one hurdle was an interview with the eminent mathematician G. F. J. Temple, who Salam knew would quiz him about the work of the British cosmologist Arthur Eddington. Suitably prepared, Salam said that he had been too impressed by Eddington's work to be able to read it objectively.[20]

The chair was as Professor of Applied Mathematics in Imperial's Mathematics Department, but Salam knew there was a good chance that he would be transferred to the Physics Department when its new buildings were ready. Lecturers at a prestigious university like Cambridge were frequently approached with offers of professorial chairs elsewhere. At that time, professorial chairs at Cambridge were rare and a lectureship was the normal 'career' post. Many lecturers turned down job offers from outside, preferring to enjoy the unique Cambridge atmosphere and reputation. However, Salam's offer was different. Although initially it looked like he would be going to a relatively minor mathematics school, he knew that soon he would become part of a major new physics department, with an international outlook. Cambridge did what it could to dissuade him from going to what it called 'Blacksmiths' College'. He was tempted by Nevill Mott with additional responsibilities, such as deputy editorship of *Philosophical Magazine*, which would bring some extra money, possibly of interest, and a personal supply of sherry, which certainly did not impress Salam, a strictly teetotal Muslim. Another Blackett recruit for Imperial was Dennis Gabor, as Professor of Applied Electron Physics, and who went on to win the 1971 Nobel Physics Prize for his invention and development of the holograph.

In London, Salam settled in leafy Campion Road, Putney, in the same south-west quadrant of London as Imperial College, and so within easy reach. It was also convenient for Heathrow Airport, but, more importantly for Salam, within walking distance of the Ahmadi Mosque in Southfields. London had been the first overseas branch of the Ahmadi movement. The foundation stone of the mosque, one of the first in Britain, had been laid by the second Ahmadi Khalif, Mirza Bashiruddin, in October 1924, and the building opened in 1926. Initially Salam lived in the house with his father, who came to London for a cataract operation. His brother Abdul Majid, training in Peterborough, was also a visitor. Salam's wife Amtul Hafeez had returned from Cambridge to Pakistan in November 1956, where their third daughter, Bushra, was born in December. Amtul Hafeez and her daughters moved to Putney in 1958.

In January 1957, Salam became Professor of Applied Mathematics at Imperial. He was the first Pakistani and only the second man from South Asia to have a professorial chair at a British university, the first having been Sarvepalli Radhakrishnan, as Professor of Eastern Religions and Ethics at Oxford in 1936, and who in 1962 became India's second President. Salam's inaugural professorial lecture was on 14 May 1957, where he was introduced by Patrick Blackett. Although Blackett was professor of physics, not mathematics, at Imperial, he took the chair at the meeting as Dean of the Royal College of Science. In his lecture, Salam pointed out to the new implications of left–right reflection symmetry and zero mass neutrinos, recounting how Pauli had tried to block the suggestion.

Mathematics at Imperial was housed in italianate buildings that had been the first home of the Royal College of Science and that also sheltered part of the Victoria and Albert Museum – a major London cultural centre – and the Royal College of Art. The rooms were cramped: large ones had been divided by thin partitions through which every sound could be heard. But from these modest beginnings came one of the most dynamic groups in British theoretical physics research. Soon Salam was joined at Imperial by Paul Matthews, who moved from Birmingham. Riazuddin, who Salam had taught in Lahore and who had followed him to Cambridge, had not completed his PhD studies by the time Salam transferred to London, but packed his cases again in 1957 and moved to London, where he was allowed to complete his doctorate under Salam's supervision. With his British education complete, Riazuddin returned to Pakistan in September 1959 to join Punjab University. Riazuddin would later found a theoretical physics group at Pakistan's new university in Islamabad, populated largely by students who had learned their trade with Salam in London. Almost immediately after Riazuddin left Imperial College, his twin brother Fayyazuddin arrived to begin PhD work, this time under Paul Matthews, staying until 1962, when he returned to Pakistan.

In their first work together at Imperial, Salam and Matthews put their loyalty to field theory to one side and used the new mathematical techniques of dispersion relations to investigate the left–right parity of several newly discovered particles. Early visitors to the group at Imperial included Oskar Klein, Victor Weisskopf, John Wheeler, who had been Feynman's teacher, and of course Wolfgang Pauli. In 1959, at the age of 33, Salam was elected to the prestigious Royal Society, the select British academy of science established in 1662 by Christopher Wren, Robert Hooke and their contemporaries, and as its youngest fellow[21]. That year, Tom Kibble joined the Applied Mathematics group at Imperial. A student at Imperial during this time was Ray Streater, later to join the group. He relates how in 1958, a fat document had come from Birmingham, a thesis by Stanley Mandelstam containing ideas on dispersion relations that kept a lot of people busy for several years, but now have been largely forgotten. Salam passed the package to Streater with the message 'this seems to be important'. After reading it, Streater immediately wished he had thought of the idea, but comforted himself with the thought that the paper was based on a conjecture, not a proved result. After a few days, Salam asked Streater how he was getting on. Streater tried to disguise his disinterest. 'Should I try to prove Mandelstam's relations?' he asked. Salam replied, 'you can try to DISPROVE them if you like; if you don't want the document, let me have it back'. Soon he gave Streater another paper, whose first page used an unfamiliar mathematical concept. Streater hesitated for a few days before asking Salam to explain it. But they were working in a department of mathematics. The impatient Salam exploded: 'these things can be found out! If you don't want the problem ...'[22] To lure distinguished speakers to Imperial, Salam and Matthews resorted to imaginative strategies, sometimes picking them up at the airport and arranging entertainment in London to suit their individual tastes.

Salam transferred from Imperial's Mathematics department to Physics in 1960, becoming the first occupant of a new Chair of Theoretical Physics. In February 1980, when Salam gave his first talk at Imperial College after receiving his Nobel prize, he was introduced by the Rector of the College, the distinguished nuclear physicist Brian Flowers (by then Lord Flowers) who said 'I think I can now reveal a closely guarded secret. The Chair [of Theoretical Physics] was originally offered to me. I turned it down, to the lasting benefit of the College!'[23] Flowers became Professor of Theoretical Physics at Manchester in 1958 and was promoted to Head of its Physics Department in 1961. While there is no doubt that Flowers would have been an admirable Head of Theoretical Physics at Imperial, the group would not have gone on to attain the same international status that Salam was able to achieve. After leaving Imperial's Mathematics Department, Salam returned there from time to time to teach quantum mechanics to undergraduates. These courses followed Dirac's traditional formalism and were in the Salam tradition of catering for more ambitious students. In one lecture, he looked round in surprise when laughter erupted after he wrote on the board that it was important to 'cheque' a certain result. It was one of his rare mistakes in English.

In his 1957 inaugural lecture, after summarizing the current status of elementary particles and the problems still to be solved, Salam had added 'how deeply privileged our generation is to have been presented with this fascinating challenge.....stepping stones to an inner harmony, a deep pervading symmetry which we shall discover.' Symmetry was to be the keyword for the next step.

REFERENCES

1. There is a third, gamma radiation, found to be high-energy electromagnetic radiation
2. Woit, P., *Not even wrong*, (New York, NY, Random House, 2006)
3. Cropper, W., *Great physicists* (Oxford, OUP, 2001)
4. Enz, ,C., *No time to be brief – A scientific biography of Wolfgang Pauli*, (Oxford, OUP, 2002)
5. Pais, A., *Inward bound* (Oxford, OUP 1986) Chap. 14
6. Reines, F., *The detection of Pauli's neutrino* in *History of original ideas and basic discoveries in particle physics*, Newman, H., Ypsilantis, T., (ed.) (New York, NY, Plenum, 1996)
7. According to Reines, (1996) but this is refuted in Enz (2002) 489
8. Salam, A., *Gauge unification of fundamental forces* Rev. Modern Physics. **52**, 525–38 (1980)
9. Enz (2002) 254
10. letter to V. Weisskopf, see Enz (2002)
11. http://www.opticsjournal.com/jcward.pdf
12. Letter from W. Pauli to V. Weisskopf, 27 January 1957 (Pauli CERN Archives). Reproduced and translated in Kronig, R., and Weisskopf, V. (ed.), *Collected scientific papers of Wolfgang Pauli*, (New York, NY, Wiley-Interscience, 1964)
13. according to Herwig Schopper, who organized the meeting
14. R. Peierls to Salam, Trieste archives B268
15. see Salam (1980)
16. Salam, A., *A life of physics*, in Cerderia, H. A., Lundqvist, S. O., (ed.) *Frontiers of physics, high technology and mathematics* (Singapore, World Scientific, 1990)
17. Pauli archives at CERN
18. Salam, A., Endeavour, 17, 97 (1958).
19. In 1930, the prize had gone to Chandrasekhara Venkata Raman in India, and in 1949 to Hideki Yukawa in Japan.
20. Vauthier, J. *Abdus Salam, un physicien* (Paris, Beauchesne, 1990)
21. Although the first Fellow of the Royal Society from the Indian subcontinent is widely assumed to have been Srinivasa Ramanujan, it was in fact Ardaseer Cursetjee (1808–77), a Bombay marine engineer, elected in 1841.
22. http://www.mth.kcl.ac.uk/~streater/salam.html
23. Imperial College archives, file KP/13/1/3

9

The arrogant theory

When Abdus Salam returned to Cambridge from Pakistan in 1954, subnuclear physics was not as chaotic as Lahore's anti-Ahmadi riots, but he knew that the scientific optimism of 1951 had evaporated. To further cloud the immediate outlook, Robert Hofstadter at Stanford, California, shone beams of electrons on nuclei and found that the constituent neutrons and protons were no longer the pinpoints of nuclear matter that scientists had thought them to be. Their definite size and shape now suggested they were not indivisible. Atom, nucleus, nucleon, . . . the depths of matter looked limitless. When would subnuclear bedrock be found?

Awaiting an answer, the post-war subnuclear physics juggernaut rolled on. New and progressively bigger particle accelerators were built, equipped with ingenious detectors to record the tracks left by wraith-like unstable particles existing only for a tiny fraction of a second. With these machines, major discoveries seemed to be made almost every week. In the decade from 1949–59, the number of known subnuclear species inflated. Instead of just the proton and neutron, and the π meson (or pion) with its three possible electric charge versions (positive, negative and neutral), the subnuclear rollcall grew to about fifty. A team at the University of California, Berkeley, periodically published updated lists, and subnuclear physicists carried pocket-sized editions next to their driving licences. With so much exotica, attention turned away from the mechanics of nuclear forces to simply understanding why there were so many different particles. What lay behind this riot of multiplicity?

To help classify them, physicists kept careful tally of the different attributes – quantum numbers – of these exotic particles. They had an electric charge, and in any particle interaction these electric charges got shuffled around, but the total charge had to be conserved – for example a neutral particle decayed into a charged pair; one positive, the other negative. Other quantum numbers were their spin (rotation about an internal axis), parity (left–right handedness), and a new charge-like quantity, which in 1953 the US physicist Murray Gell-Mann called 'strangeness', a quantum indicator of exoticness. The familiar nucleons and pions had strangeness zero, but exotic particles warranted strangenesses of ±1, ±2, ... Strangeness was the first example of a bizarre new physics vocabulary. Older, more staid physicists preferred the sophistication of abstract words like 'parity' or 'renormalization' and rejected Gell-Mann's contrived banality, preferring instead to say 'hypercharge', but Gell-Mann's choice stuck[1]. Why shouldn't strange particles carry 'strangeness'? It was a throwback to earlier centuries when basic words like 'field' and 'gravity' had been borrowed from everyday usage and endowed with deeper meanings.

The first strange particles had been spotted in the late 1940s in photographs of stubby tracks left by cosmic particles. Later, physicists learned how to fashion strange particles into beams and thereby manufacture even stranger ones. When the beams smashed into targets, the strangeness initially had to be conserved, like electric charge. But strange particles had their own versions of beta decay, decaying into less strange particles, shedding strangeness in a way that appeared to follow certain mysterious rules.

It was natural to think of the proton and the neutron being close relatives, the simplest of isotopes, distinguished by their electric charge. Just as the quantum electron can only spin in one of two ways – clockwise or anticlockwise – so a generic particle, the nucleon, could be viewed as 'spinning' in some abstract isotopic space, where clockwise and anticlockwise translate into a proton and a neutron, respectively. (Instead of clockwise or anticlockwise rotation, physicists prefer to think in terms of the direction of the axis of this spin, pointing either 'up' or 'down'.)

At the end of his 1957 inaugural professorial lecture at Imperial College, London, Salam had intimated that some kind of symmetry must be the key to understanding all these particles. Such symmetry would be revealed in some wider scheme, extending the abstract but compelling idea of the isotopic spin that distinguished protons from neutrons. Already the hunt was on for patterns that reflected such symmetry. In 1955, while still at Cambridge, Salam and his research student John Polkinghorne put forward ideas based on isotopic spin in four dimensions. Some of its implications were similar to a scheme dreamed up by Gell-Mann and presented at major physics meetings in 1954.

Although the wider picture was still fuzzy, one segment of it had come into sharp focus. One unstable strange particle, which had been awarded the Greek letter lambda, resembled a heavier cousin of the proton and the neutron. Here was a subnuclear trio, with the lambda's strangeness endowing it with extra mass. To accommodate particles that can have strangeness as well as electric charge needed a mathematical model that extended the binary isotopic spin up/down description of nucleons into a three-dimensional picture. Physicists knew by now that the proton and the neutron, and presumably the lambda too, had shape and structure, so any triplicity of building blocks should not be these particles themselves. Nevertheless, Shoichi Sakata in Japan proposed a theory based on the physical proton, neutron and lambda. Composite particles were composite because they were made up of each other. The idea beguiled Salam.

In their search for symmetry, physicists were straying into what was for them unfamiliar territory. For mathematicians, a particular symmetry is the trademark of an underlying 'group'. Regular shapes, such as a triangle, a square, or a pentagon, remain unchanged under certain rotations – 120° for an equilateral triangle, 90° for a square, 72° for a pentagon. Each set of possible such rotations is the symmetry group of a particular geometry. Such striking symmetries are termed discrete – their rotations either happen or they do not. More subtle are symmetries that are instead continuous – a sphere remains a sphere no matter how much or how little it is rotated in any direction. The mathematics of such continuous groups had been developed by the Norwegian mathematician Sophus Lie (1842–1899) and Elie Cartan (1869–1951) in France. The relevance of group theory for modern physics had been pointed out by Hermann Weyl in his book *Gruppentheorie und Quantenmechanik*, first published in 1928, and the message underlined by Eugene Wigner in 1931. Unfortunately, Weyl's was a difficult, abstract book, and the importance of its message evaporated as time passed. One who persevered with it was the Italian-born Israeli mathematician Yoel (Giulio) Racah, who in 1951 had given a course of lectures on group theory at the Institute for Advanced Study in Princeton. Although many leading contemporary physicists knew of the lectures and may even have gone to them, few seemed to have taken any note. As a 21-year old parvenu, Gell-Mann had attended the lectures, but had not immediately seen their relevance. There are no notes from Racah's lectures in Salam's papers. While physicists struggled with crude rotational symmetries, Weyl's message as relayed by Wigner and Racah went unheeded.

Salam and Polkinghorne's 1955 paper had used the unfamiliar mathematics of the rotation group in four dimensions. They were following signposts erected the previous year by Abraham Pais, a gifted Dutchman who emigrated to the USA after traumatic wartime experiences, and who had attended Racah's lectures at Princeton. As a mathematics undergraduate at Cambridge, Salam had learned something about group theory, but was handicapped by not having heard Racah. Polkinghorne, in deep mathematical water, consulted his colleague Ronald Shaw, whose help was acknowledged.

Two theorists did struggle through the mathematical thicket of group theory: they were Murray Gell-Mann, who had landed up at the California Institute of Technology (Caltech), Pasadena, and Yuval Ne'eman in Salam's group at Imperial College, London. These men had many things in common, and would eventually work together, but had initially chosen to follow very different careers. Their names were curiously similar: both were polymaths with eidetic memories and a gift for languages. After his invention of the strangeness label, Gell-Mann went on to attack many of the basic problems of subnuclear physics. Richard Feynman was to say of him 'The development during the past twenty years of our knowledge of fundamental physics contains not one fruitful idea that does not carry his name'. In 1960, Gell-Mann was struggling, as Salam and Polkinghorne had done in 1955, to classify the haphazard plethora of exotic subnuclear particles into patterns that reflected their different properties. Nothing seemed to make sense. Feynman, whose office was just down the corridor at Caltech, suggested that if Gell-Mann was trying to do group theory, he should consult a mathematician. Gell-Mann did so and realized that he had been painfully rediscovering what mathematicians already knew, and that he should have listened more to Racah. As his blindfold dropped away, Gell-Mann discovered subtle symmetries that generated patterns, regular shapes like stars or hexagons, all subdivided into triangles. Each apex of every triangle could be assigned to an exotic particle, providing a possible subnuclear seating plan. (In the early 1960s, as experiments continually stumbled across exotic particles, some of their quantum numbers were wrongly assigned by the initial experiments, confusing the seating arrangements.)

Yuval Ne'eman was born in Tel-Aviv in 1925 and initially studied engineering, the family business. But as he reached the age when many people start to think about their own future, his homeland – the British Mandate of Palestine – itself had an uncertain future, and he joined *Haganah*, the Jewish underground resistance movement. When Israel became a nation in 1948, *Haganah* provided the nucleus of the new nation's army. After commanding an infantry brigade in Israel's 1948 war of independence, Ne'eman remained a professional soldier, foraying into the murky world of intelligence and covert operations. After the 1956 Suez crisis, he wanted to take time out from the military and pursue another ambition – to study physics, and try to exploit Einstein's ideas for unifying gravity with electromagnetism. But Moshe Dayan, Chief of Staff of the Israeli Army, had other plans for him. Now a colonel, Ne'eman arrived in London as defence attaché to the Israeli Embassy. He was hoping for a quiet life, so that he would have plenty of time to study physics, but soon the Middle East was again in turmoil. After the Suez crisis, Egyptian President Nasser became an inspiration to other Arab nations. In 1958, a revolution in Iraq overthrew the monarchy. The British feared that Jordan would follow, and wanted to overfly Israel to bring support for Jordan's King Hussein.

When not negotiating the trade-off between overflying rights and British armour for the Israeli army and submarines for its navy, Ne'eman had initially tried to follow a graduate course in cosmology and relativity at London's King's College, but in those hectic days found it difficult to combine the two. Imperial College also did physics, and instead of a cross-town commute, was only five minutes' walk from the Israeli Embassy. Reading the Imperial prospectus, Ne'eman learned that Salam's group did field theory, albeit of a very different kind to Einstein. Ne'eman went to Salam and presented his only accreditation, a letter from General Moshe Dayan. Salam was amused.

Salam and Ne'eman were the same age, 32, young for a professor, but old for a research student. To his credit, Salam took on Ne'eman for a short-term research project. Later Ne'eman said: 'From the first day he behaved as a gentleman, giving me a chance. It would not have been hard for him to find some excuse and send me elsewhere.'[2] After the bitter war between Israel and its Arab neighbours in 1948, the appearance of a high-ranking officer of the Israeli army could easily have displeased a Muslim professor. After official duties, Ne'eman would change out of his uniform, sometimes in his car, before appearing at Imperial College.

During his first year, his military duties conflicted with the lectures he was supposed to follow, but he was impressed by Salam's talks on symmetries and group theory, and in May 1960 began full-time research on leave from the army with a one-year scholarship from the Israeli government. The first thing was to ask Salam for a research problem. Normally such projects are supposed to be difficult enough to occupy several years, but nevertheless tractable, as research students have to be sure of having something to show for their efforts. But Ne'eman had only one year. Salam decided to trade time for something more speculative, and told Ne'eman to look into the mathematical foundations of group theory. 'You are embarking on a highly speculative quest,' Salam told him nervously. 'You tell me you have a total of one year, so I have tried to devise for you something you can finish in a year.'[3]

Salam was himself exploring classification schemes for particles, but gave Ne'eman free rein. 'Do not stay with the little group theory I know, which is what I taught you, do it in depth', he urged the Israeli officer[4]. Adrift in deep mathematical water, Ne'eman's first lifebelt was a translation he discovered in the British Museum Library of an obscure Russian mathematical paper by the Soviet mathematician E. B. Dynkin, which reclassified the symmetry groups explored by Cartan. After surveying the catalogue of symmetry schemes, Ne'eman was left with just two possibilities for the subnuclear particle seating plan. The mathematicians had called them G_2 and SU(3). Ne'eman initially had been attracted to G_2 because it had a Star of David symmetry pattern: SU(3) had meanwhile also been discovered by Gell-Mann.

SU(3) looked familiar because physicists knew its little brother, SU(2), the symmetry of isotopic spin. One characteristic SU(3) symmetry pattern is a hexagon. As well as the particles assigned to the six apexes of the figure, two more can sit at its centre, an arrangement that Gell-Mann, with his thirst for innovative nomenclature, had called the 'Eightfold Way', a traditional exhortation to Buddhist monks.

Intrigued but unconvinced by Ne'eman's discovery, Salam suggested that they should write a paper together, but nothing emerged from Salam. Tired of waiting, Ne'eman reminded Salam, who said that he had now changed his mind about a joint paper, but added a suggestion that Ne'eman duly took up. Ne'eman naturally wanted to publish his discovery as rapidly as possible, and as an unknown student tried to bypass the Imperial College typing pool by using the Israeli embassy secretaries instead. Unfortunately the embassy was not used to typing physics papers for publication and used the wrong format[5]. The paper was immediately rejected by the journal *Nuclear Physics* on these

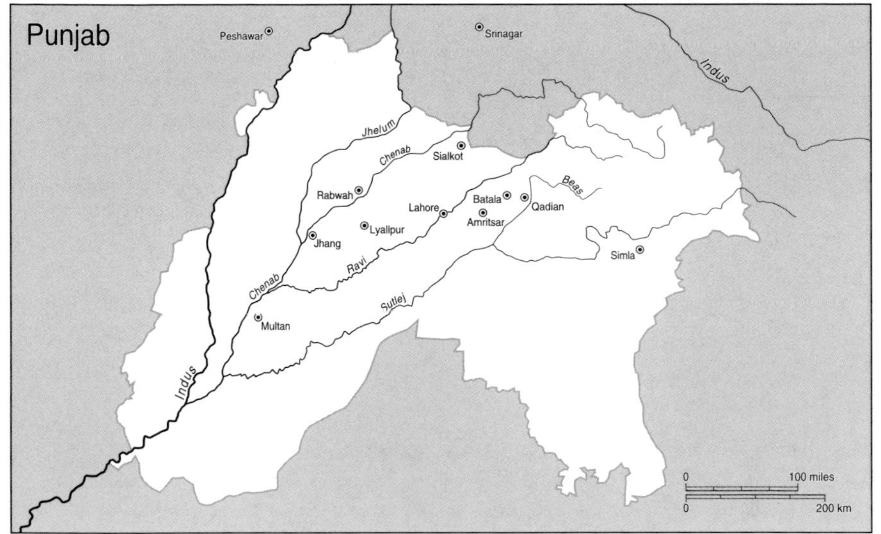

1. The Punjab in British India. The name means 'Land of Five Waters', the Jhelum, Chenab, Ravi, Sutlej and Beas, tributaries of the mighty Indus. In 1947, the province was split between the new nations of India and Pakistan.

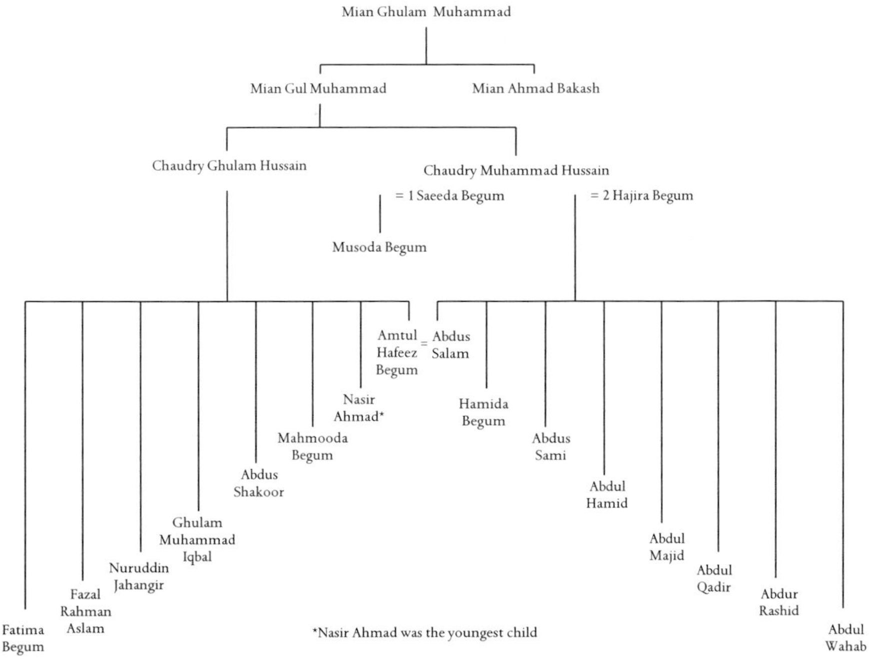

- Mian Ghulam Muhammad
  - Mian Gul Muhammad
    - Chaudry Ghulam Hussain
      - Fatima Begum
      - Fazal Rahman Aslam
      - Nuruddin Jahangir
      - Ghulam Muhammad Iqbal
      - Abdus Shakoor
      - Mahmooda Begum
      - Nasir Ahmad*
      - Amtul Hafeez Begum = Abdus Salam
    - Chaudry Muhammad Hussain
      - = 1 Saeeda Begum
        - Musoda Begum
      - = 2 Hajira Begum
        - Abdus Salam
        - Hamida Begum
        - Abdus Sami
        - Abdul Hamid
        - Abdul Majid
        - Abdul Qadir
        - Abdur Rashid
        - Abdul Wahab
  - Mian Ahmad Bakash

*Nasir Ahmad was the youngest child

2. Abdus Salam's family tree. His uncle and father-in-law Chaudry Ghulam Hussain was the first in the family to join the Ahmadi sect of Islam.

3. Hazrat Mirza Ghulam Ahmad (1835–1908), the Promised Messiah of the Ahmadi sect of Islam, to which Salam's family belonged.

(Photo courtesy of the Ahmadi Muslim Community of the UK)

4. Chaudry Muhammad Hussain, Abdus Salam's father. Convinced of his son's prowess, he firmly guided Salam's early studies at every stage, and taught him to love Islam. Salam had enormous respect for him.

5. Punjab education: Jhang elementary school, which Salam attended from the age of six.

M. ABDUS SALAM

Copy of letter from Mr. Abdus Salam who stood first and smashed the Punjab University Record by securing 765 marks out of 850 in the M. S. L. C. Examination 1940.)

To

THE MANAGER,

PUNJAB KITAB GHAR (Regd.)

19, Mohan Lal Road, LAHORE.

Dear Sir,

I have a great pleasure in thanking you for the useful books, you publish for the Matriculation students. Your books are really valuable and I have no hesitation in saying that the following books are of outstanding merit, and are most useful for the boys.

Solved Exercises in Geometry by Dharam Dutt.

Modern English Grammar and Composition.

Yours truly,

6. In 1940, Abdus Salam, age 14, demolished Punjab University's Matriculation and School Leaving Certificate (MSLC) examination. With such excellent marks over a wide range of subjects, it was not clear which direction his future studies should take.

7. The impressive Main Hall of Government College, Lahore (now Government College University) has a central, soaring nave surrounded by double-storied aisles. Salam was a student at Government College from 1942–6 and Professor of Mathematics from 1951–4. In 1997 the building was renamed Salam Hall.

(Photo M. Nadeem Malik, Department of Physics, GC University, Lahore)

8. Victorian Gothic in the Punjab: the outside of Salam Hall at Government College University, Lahore. (Photo M. Nadeem Malik, Department of Physics, GC University, Lahore)

9. At New Court, St. John's College, Cambridge in 1946, the ornate neo-Gothic architecture could have reminded the travel-weary Salam of the Mughal splendour of Lahore. (By permission of the Master and Fellows of St John's College, Cambridge)

10. Fred Hoyle in the very early 1950s in his study at St John's College, Cambridge. Just a few years before, Salam was tutored in undergraduate mathematics by Hoyle in this room. It was Hoyle who convinced Salam to switch from mathematics to physics.

(Credit: St John's College Cambridge, Hoyle archive)

11. Paul Matthews (1919–87, right) was Abdus Salam's first research collaborator and became his lifelong friend. This photo was taken in 1963 at Imperial College, London, where they were both Professors of Theoretical Physics.

(Photo: Blackett Laboratory, Imperial College, London)

12. Sir Muhammad Zafrullah Khan (1893–1985) was a great influence on Abdus Salam, and introduced him to the world of the United Nations and its agencies. Doyen of the Ahmadi community and a member of the Supreme Court of British India, Zafrullah Khan was Pakistan's first Foreign Minister, and subsequently became the nation's permanent representative at the UN. Distanced, like Salam, from Pakistan by his Ahmadi belief, Zafrullah Khan moved to the International Court of Justice in the Hague, where he was its President from 1970–73. He is seen here in his ICJ robe and jabot. (Photo courtesy of The International Court of Justice)

13. Trieste Institute co-founder Paolo Budinich (*right*) with quantum pioneer Wolfgang Pauli in Zurich in 1954. Young physicists feared and revered the caustic Pauli. In 1956, rebuked by Pauli, Salam held back publication of his explanation that the neutrino is responsible for left-right asymmetry in nuclear beta decay, a decision that both Salam and Pauli later regretted. (Photo courtesy of Paolo Budinich)

14. It was with John Ward that Salam beat a path towards the unification of electromagnetic and weak nuclear interactions. Ward said '[Salam] and I were old friends, despite the fact that our temperaments were directly opposite'. This photo was taken in April 1964 during a countryside tour arranged by Richard Dalitz of Oxford. (Credit: J. D. Jackson, courtesy AIP Emilio Segrè Visual Archives, Jackson collection)

15. In January 1965, at a meeting on 'Symmetry Principles at High Energy', held at the University of Miami, Coral Gables, a smug-looking Salam presented his latest theory—grandly entitled 'The Covariant Theory of Strong Interaction Symmetries'. The session chairman was J. Robert Oppenheimer (*left*), who at the time was also a key member of the Scientific Council for Salam's Trieste Centre. The ebullient Salam was convinced that the theory had 'solved everything', and impressed the audience by pulling results from his new theoretical hat. Oppenheimer was not convinced, Salam's theory was soon found wanting, and sank virtually without trace.
(ICTP Photo Archives)

16. Salam's first visit to Trieste was for a physics meeting in June 1960 organized by Paolo Budinich at the castelletto of Archduke Maximilian's Miramare estate. This sparked the Salam-Budinich collaboration that led to the establishment of the International Centre for Theoretical Physics. This is one of the rare occasions that Salam (*front row, right*) appears without a jacket. Budinich is to his right. Also in the front row (*jacket, open-necked shirt*) is Jacques Prentki, an early Salam research collaborator. (ICTP Photo Archives)

17. Vienna, 1964. An early meeting of the Scientific Council of the International Centre for Theoretical Physics under the auspices of the International Atomic Energy Authority (IAEA). Left to right: Salam, Paolo Budinich, L. Liebermann (IAEA), A. Sanielevici (IAEA), H. Seligman (IAEA), and J. Robert Oppenheimer, Director of the Institute for Advanced Studies, Princeton, and formerly Scientific Director of the Manhattan wartime atomic bomb project. The Salam-Oppenheimer symbiosis had begun when Salam worked at Princeton in 1951. (ICTP Photo Archives)

18. At Duino Castle, Trieste, front row: Sigvard Eklund, Secretary General of the International Atomic Energy Authority (IAEA), Prince Raimondo of Torre e Tasso, and Salam. Eklund was a staunch supporter of Salam's international plans, and the Prince was a major local benefactor, donating the land that would become the future home of the International Centre for Theoretical Physics. (ICTP Photo Archives)

19. From 1965–8, the first home of the International Centre for Theoretical Physics was on an upper floor of this building in Trieste's central Piazza Oberdan.
(ICTP Photo Archives)

20. 1968 Flags hoisted at the formal opening of the new building of the International Centre for Theoretical Physics, Miramare, Trieste. (ICTP Photo Archives)

21. Aerial view of Miramare, seven kilometres north of Trieste, showing Archduke Maximilian's castle overlooking the Adriatic. (ICTP Photo Archives)

22. The meeting that marked the opening of the new building of the International Centre for Theoretical Physics in 1968 included autobiographical sketches by key figures, including quantum mechanics pioneer Paul Dirac, who according to Salam 'represented the highest reaches of personal integrity of any human being I have ever met'. Dirac's lectures on quantum mechanics were a turning point in Salam's mathematics undergraduate career at Cambridge. In the early 1950s, Salam joined Dirac as a fellow of St. John's College.

(ICTP Photo Archives)

23. When the recipients of Nobel Prizes collect their awards from the King of Sweden, men usually wear formal dress and a white tie. On 10 December 1979, Abdus Salam's traditional Punjabi dress, with a turban and khusa shoes, epitomized the huge leap of his life's accomplishment. (PA Photos)

24. Seifallah Randjbar-Daemi, Salam's longtime research collaborator at Trieste, with him at Louise Johnson's house in Oxford in October 1996, just one month before Salam died.

technicalities before even being read. Meanwhile, a preprint version of Gell-Mann's SU(3) ideas arrived on Salam's desk. Salam knew Gell-Mann's reputation, and on receiving the draft immediately upgraded his opinion of Ne'eman and of SU(3), and climbed aboard the SU(3) bandwagon. Salam and Matthews had been so enthusiastic about Sakata's suggestion based on the triplet of the proton, the neutron and the strangeness-carrying lambda that they completely overlooked the implications of Ne'eman's work until the arrival of Gell-Mann's paper.

Although proud of having 'discovered' Ne'eman and guiding his research, Salam must have regretted never collaborating with him. He did not get another chance because Ne'eman returned to Israel in 1961 to become scientific director of the country's nuclear research programme, while maintaining his interest in SU(3). In 1962, a physics conference at CERN focused on the latest news from the exotic particle front. In SU(3), another of the available symmetry patterns was a large triangle whose grid could accommodate ten particles. However, three of these sites did not correspond to anything that was known, and the decuplet had been discounted as a meaningful physics item. But the 1962 meeting brought news of two new particles that exactly fitted into the triangular diagram, which now only had one vacancy. At the end of the talk, the chairman invited questions from the floor. Both Gell-Mann and Ne'eman raised their hand, but it was the better-known figure of Gell-Mann who strode to the blackboard and dramatically predicted what the lone missing particle in the decuplet should look like, calling it the 'omega-minus'. In the audience other physicists took note and set about looking for it when they got home. In early 1964 came news that it had finally been found. The entire life history of the ephemeral particle had been captured in a single photograph showing its creation and the whiskery tracks of its decay, dramatic proof of the validity of a theory that few people could yet understand. Reporting the discovery, the *Times* of London said 'Three years ago, Dr Gell-Mann and another scientist, working independently, proposed to classify the well-known particles in groups.' Salam appeared as an expert witness on BBC television news, but his replies to questions were far from crystal-clear to most viewers.

With SU(3) now established, Gell-Mann realized that he had to explain where its intrinsic triplicity came from, and in 1963 launched his idea of 'quarks', an invisible triplet of building blocks from which all the other subnuclear particles could be built up, a sharp mathematical image of the mongrel Sakata model. 'Quark' (pronounced 'quork' by Gell-Mann but by few others) was a nonsense word that became a turning point in science. The individual quarks were christened less flamboyantly: the pair that made up protons and neutrons were labelled 'up' and 'down', reflecting the two different orientations of isotopic spin; with the third 'strange' quark carrying Gell-Mann's quantum label. Other physicists found in these quarks a simple idea that could make equally simple predictions, but such quark games were too simple for Gell-Mann. Now working backwards from SU(3) and in collaboration with Sheldon Glashow, he showed that physics needed special types of mathematical groups, a result also obtained at Imperial by Penelope Ionides (later Rowlatt) after suggestions by Salam and Ne'eman.

The intriguing quarks had a major drawback – they carried fractional electric charges, something that was difficult to swallow, particularly for Salam. Ten years later, when he came to propose a wider theory involving quarks, he initially insisted that they had to carry integer electric charges. 'If you have all integer charges . . . you can even teach it to a child', he said[6]. The Sakata model had not suffered from any fractional charge handicap.

During Ne'eman's time at Imperial, Salam was maintaining close links with John Ward, a highly talented British researcher whose name now appears on several physics milestones. Ward was also highly nomadic, his career seeming to lurch unpredictably from one research destination to the next[7]. Despite this chaotic background, Ward's achievements were remarkable, seemingly able to solve any intellectual problem that he encountered. After a wartime degree in engineering and then mathematics at Oxford, he dabbled in theoretical quantum physics before taking up a lectureship at the University of Sydney, Australia, where he remained for just one year. Returning to research at Oxford, his research thesis was torpedoed by Rudolf Peierls, replacing Nicholas Kemmer as external examiner at the last minute. The crestfallen Ward wanted to leave academia for a job as a trainee engineer at Rolls-Royce, but was lured back by Oxford. Here, he started looking at Dyson's problems of renormalization in quantum electrodynamics, discovering in 1950 a profound theorem that became known as the 'Ward Identity'. It was at this time that Salam and Ward first met, when Ward was invited to give a talk at Cambridge. In mathematics, little is more impressive than a mathematician who has lent his name to a theorem. Ward's reputation was made. After leaving

Oxford, he went in 1951 for one year to the Institute for Advanced Study in Princeton, where he again met fellow renormalization specialists Salam and Matthews. After a year at Princeton, Ward was again confused about what to do next, and left academia for Bell Laboratories, New Jersey, to work on electron-beam techniques. But again the nomadic Ward quickly moved on, this time again to Australia, for a job at the University of Adelaide, only to change job abruptly once more, this time (1955) at the UK Atomic Weapons Research Establishment at Aldermaston, before returning for a short period to the US for industrial electron work, again for just a year. In 1956 he moved back into academia, this time at the University of Maryland, just in time to hear the news that parity was violated in weak interactions.

Salam's fruitful collaboration with Ward had to fit into this erratic scenario, but it was with him that Salam was to craft the contribution to science for which he is best remembered, and went on to earn him the Nobel Prize. In his memoirs[8], Ward wrote 'My attitude towards [theory] differed strongly from other practitioners. Many seemed to regard the subject as a kind of glorified Klondike gold rush, staking their claims as best they could and keeping their cards close to their chests . . . I perversely refused to play the game, Instead, I would openly discuss the problems with anyone who was interested, and in particular of course with Abdus [Salam]. He and I were old friends, even though our temperaments were directly opposite. He would publish anything and hope for the best. I would not normally publish unless I was sure of the product. Strangely enough, he would also put my name on papers, if we had discussed the problem, without asking my permission'. After Maryland, Ward continued his peripatetic travels with short stints in Miami, Carnegie (Pittsburgh), Princeton again, Johns Hopkins (Baltimore), Wellington (New Zealand), and finally Macquarie (Sydney), where he finally settled down.

The SU(3) of Gell-Mann and Ne'eman was in fact a subset of a symmetry that had already been visited. In the 1930s, the realization that the forces between nuclear protons and neutrons were approximately the same, no matter which way the nucleons' spins or isotopic spins were arranged, had led the Hungarian-born Eugene Wigner to develop a picture of nuclear forces described by SU(4), a larger symmetry than SU(3). These ideas led to Wigner being awarded the Nobel Physics Prize in 1963, soon after the appearance of the SU(3) quark picture, a timely reminder to physicists that symmetry based on group theory was not a new idea. This earlier work now pointed the way ahead, and the idea took root of a larger SU(6) symmetry of quarks, based on a sextet made up of the three kinds of quark, each with the customary up/down spin assignments. In this picture, the eightfold way of Gell-Mann and Ne'eman was enlarged to 36, or even 56-fold possibilities. Its great success came in 1964 with the explanation of the relative magnetism of the neutron and the proton – the predicted value was –2/3, compared to the measured value of –0.685 (the minus sign means the two magnetic effects point in opposite directions).

So far Salam had been on the periphery of the SU(3) and quark game, but he nurtured an ambitious plan. In 1960, even before the days of SU(3) and quarks, he had tried with Ward to construct a theory of all particles and all interactions using rickety eight- and 16-dimensional struts containing the known particles, and bolting these together. Salam and Ward realized that the attempt had been unsuccessful, but thought nevertheless that the result might be 'of interest'[9]. After SU(3) and the introduction of the quark idea, the struts needed to build such theories became shorter and more rigid, but for Salam's objective these 'internal symmetrie symmetrics' had to be combined with classic features of particle behaviour. In the white heat that forged modern quantum mechanics from unmalleable concepts in the 1920s, Wolfgang Pauli had introduced the mathematics of spin through a set of toy two-by-two matrix constructions whose unusual algebra gave the required behaviour. The genius of Paul Dirac had then shown if an electron had to conform to Einstein's picture of relativity, the quantum dimensionality of the electron had to be increased from Pauli's two to four: eventually assigned to electron spin up, electron spin down, antielectron spin up, antielectron spin down. The discovery of antielectrons (positrons) in the early 1930s had been an impressive demonstration of how simple assumptions can have dramatic implications.

Salam's plan was to do for quarks what Dirac had done for electrons – make the theory conform to relativity. The quark triplet had to be meshed with Dirac's fourfold structure in a framework with twelve dimensions. In 1964, Salam was juggling with several slippery balls simultaneously: as well as his research, he was now overseeing the launch of his new International Centre for Theoretical Physics in Trieste, Italy. Awaiting construction of its permanent home north of the city, temporary accommodation in the town centre had become the embryo of a new physics research centre. This was home to a group of Salam's research assistants and students, most of whom had moved from Imperial College to Trieste. Two of them – Robert Delbourgo and John Strathdee – became Salam's research lieutenants in his quest to synthesize quarks and relativity.

Salam's furious inventiveness functioned best when he worked with partners who could then channel and temper his ideas. Paul Matthews had been the first to play this role. The collaboration between Salam and John Ward was more sporadic, but both these long-standing collaborations came to an end when Salam began operating from Trieste. After 1964, with the exception of invited contributions to conferences and meetings, Salam rarely signed a scientific paper alone. In these collaborations, Delbourgo and Strathdee were the first to inherit the roles previously played by Matthews and Ward.

Robert Delbourgo, like Salam, Paul Matthews and Tom Kibble, was born in British India – in Bombay in 1940. However, this was somewhat accidental: his father ran an import–export business in Aden. In 1940 this strategic port was bombed daily by the Italian Air Force, and much of the population evacuated to India. Robert was educated first in Alexandria, then in England, going on to study physics at Imperial College. Opting to do theoretical physics as a special subject in his final undergraduate year, he was impressed by Salam's lectures. The large classes at Imperial made lecturers distant figures, but Delbourgo invited Salam to a student dinner and had been impressed to see how well he could interact with students. Delbourgo embarked on postgraduate work at Imperial, sitting alongside Yuval Ne'eman. After working with Salam on field theory problems, he moved to a postdoctorate position at Madison, Wisconsin, which was on Salam's regular travel itinerary, and together they wrote a paper on renormalized electrodynamics in 1964, a subject now better known by the name of 'Schwinger–Dyson equations'. When in 1964 Salam asked if he would be interested in moving to the new centre at Trieste, Delbourgo jumped at the opportunity.

After graduating from Montreal's McGill University in 1956, John Strathdee went to Cambridge for a year, where he followed lectures by Salam, before moving to Dublin's Institute for Advanced Studies. After military service, in 1961 he enrolled as a graduate student at Imperial College, and moved with Salam to Trieste in 1964. John Strathdee, like Delbourgo, was a master at doing long, difficult calculations with extraordinary accuracy, as well as having a very good technical expertise in many areas of mathematics and theoretical physics.

The years 1964–66 were very busy for Salam. Robert Delbourgo says 'Often we wouldn't see [Salam] for a week or two while he was away politicking; the result was when we finally did see him during the brief spells in Trieste, the activity was heightened, even feverish. He wanted rapid progress, which wasn't always feasible, and got impatient, even tetchy, when things stalled. We knew that he was desperately distracted with the various international organizations and committees he had to deal with, not to mention his obligations to Imperial College. He would work non-stop and quite often carry out his research on planes, or else would want to consult John and I about some finer detail on our research, just before taking off or just after landing. It's miraculous he was so productive, given the other burdens he had to carry.'[10]

A 1964 *Physical Review* paper by Delbourgo, Salam and Strathdee, all listed as working at the Trieste centre rather than Imperial College, described an initial attempt at making SU(6) relativistic[11]. But even before this appeared in print, the trio had forged ahead and developed $\tilde{U}(12)$, where the tilde ('twiddle') over the U reflects the difficulty of representing four-dimensional space-time symmetry as rotations in four dimensions. (Later, the symmetry was called U(6,6), reflecting a quadratic invariant with six positive and six negative terms.) The eightfold way of SU(3) had now ballooned to a 364-fold way. The paper was fast-tracked for publication in the *Proceedings of the Royal Society* and appeared in January 1965[12]. Now Salam was listed as working at Imperial College, 'on leave of absence at Trieste'. Its arrogant title 'The covariant theory of strong interaction symmetries' ('covariance' in this context meant full compatibility with relativity) presaged a watershed in elementary particle theory. Startled research students looked up from their calculations when they heard Salam in the corridor proclaiming 'you can all go home now!'. The instruction was premature, as soon they had to drop what they had been doing and start doing calculations in U(6,6) instead. The U(6,6) paper concluded that the starting point for calculating strong interaction effects would be a 'trivial step'. Students were soon deep in calculations that were far from trivial.

Salam's new theory was not the only one on the market. With so much effort going into generalizing the SU(3) and SU(6) results, several other collaborations, with similar predictions, appeared on the scene at more or less the same time. A shop window for these efforts was a meeting on 'Symmetry Principles at High Energy', held at the University of Miami, Coral Gables, from 20–22 January 1965. Following a short introduction by J. Robert Oppenheimer on the importance of symmetries on basic physics, Salam took the stage and demonstrated his new results, pulling mathematical rabbits out of his U(6,6) hat. With the advantage of speaking first, he startled the audience with his extrovert presentation. However, as the meeting progressed, it became clear that others had produced similar results, albeit not as enthusiastically as Salam. Despite his outward confidence, Salam was particularly concerned about Oppenheimer's reaction to the theory. A photograph of the meeting (see Plate 16) shows Salam looking very smug at the end of his talk, while Oppenheimer, the chairman of the session, looks quizzically at him.

Back from Miami, Salam continued his U(6,6) publicity campaign. In Europe, he had the advantage of being on the scene early and, with affiliations to London and Trieste, was a 'home-grown' product. The London *Sunday Times*, the shop window of London's swinging sixties, immediately published a major article. As Salam had hoped, his new research centre at Trieste was suddenly on the physics map. Each week, the theory group at Imperial College had a technical seminar, usually given by an invited speaker. In January 1965 one of these seminars was scheduled to be given by Salam, still with the ambitious title 'The Covariant Theory'. After all the media publicity, students and staff from all over Imperial College turned up, and could not get into the small theory conference room. The talk was hurriedly rescheduled for the main undergraduate lecture theatre. Even then, the place was packed. Salam came in, looked astonished, then said he would spend five minutes giving a general background, after which most people could leave, he suggested. After this introduction, largely incomprehensible, he sternly told the remaining audience that there was no point in taking up theoretical physics as a career because it was now finished. People were both confused and amused. For the Imperial research students, physics was far from finished, for at the end of the seminar, calculation assignments were handed out like presents on Christmas morning.

Despite Salam's initial optimism for his gleaming new theory, the results from these calculations did not agree with experiment. By the summer of 1965, attention turned to patching up 'The Covariant Theory', for it had revealed a major shortcoming. Quantum mechanics, the calculus of the microworld, does not deal with exact cause and effect, yielding instead probabilities, and it is the average of these microscopic statistics that gives our everyday experience. However, in calculating all these probabilities, one thing is certain – something has to happen. In mathematical terms, the sum of all possibilities has to add up to 100%, a requirement called 'unitarity' in the trade. Despite all its inbuilt sophistication, which ensured that subtle rules forbidding certain reactions emerged from the calculations as if by magic, U(6,6) did not guarantee unitarity. Other theorists started analysing the basic mathematical structure of the theory, and the final nail in the coffin came in 1967 with a powerful 'no-go' theorem by Sidney Coleman and Jeffrey Mandula. Fortunately, no student had heeded Salam's January 1965 exhortation to go home. Salam's own enthusiasm for the theory quickly evaporated, and his later autobiographical talks, such as the Nobel lecture, hardly mention U(6,6) at all. However, its powerful formalism was resurrected much later as a calculational tool for theories restricted to heavy quarks.

For physics, the Delbourgo–Salam–Strathdee collaboration forged at Trieste was to endure for several more years, sometimes augmented by Pakistani physicist Muneer Rashid, later to work at Islamabad with the theory group founded by Riazuddin. Although continuing to collaborate, Delbourgo, now working from Imperial College, wanted to strike out on his own and eventually emigrated to Tasmania, Australia. However, John Strathdee remained at Trieste for 32 years, providing the perseverance and discipline that Salam's whimsical brilliance often needed. Their collaboration lasted from 1965 to 1993. Typically Salam would arrive at Trieste for a few days, bombard Strathdee with his latest ideas and then depart for a couple of weeks, leaving Strathdee to sort out the good ideas from the bad, and see if he could implement them. Although Strathdee kept himself very much in the background, he was to play an important role in Salam's later work.

But even at the height of the U(6,6) euphoria in 1965, and with his Trieste physics institute emerging as a major new research centre, Salam had not forgotten another physics goal. This was to be the biggest one of all.

REFERENCES

1. Hypercharge was adopted later for theories that brought in weak interactions
2. Y. Ne'eman, e-mail 7 March 2006
3. Y. Ne'eman, e-mail 7 March 2006

4. Ne'eman, Y., *The three-quark picture*, in *The particle century*, Fraser, G., (ed.) (Bristol, IOP Publishing, 1998)
Ne'eman, Y., *The classification and structure of hadrons* in *From pions to quarks*, Brown, L., Dresden, M., Hoddeson, L., (ed.) (Cambridge, CUP, 1989.
5. Johnson, G., *Strange beauty* (London, Jonathan Cape, 1999)
6. Crease, R., Mann, C., *The second creation, makers of the revolution in twentieth-century physics* (New York, Macmillan, 1986)
7. http://www.opticsjournal.com/JCWard.pdf
8. http://www.opticsjournal.com/JCWard.pdf
9. Salam, A., Ward, J. C., *On a gauge theory of elementary interactions*, Nuovo Cimeno, 19, 165–70, 1961
10. R. Delbourgo, to author, 22/4/06
11. Delbourgo, R., Salam, A., Strathdee, J., *Relativistic structure of SU(6)*, Physical Review, 138, B420–3 (1965)
12. Salam, A., Delbourgo, R., Strathdee, J.,*The covariant theory of strong interaction symmetries*, Proceedings of the Royal Society A285, 146–58, 1965

10

Uniting nations of science

The Second World War changed the world in many ways. It was one of the most destructive wars in human history, with more than 46 million victims – soldiers and civilians. It brought the Holocaust and the Atomic Bomb. It also galvanized science. Nuclear fission, radar and computers were all harnessed to the war effort, but to do so meant discovering a new way of making discoveries. Before the Second World War, science used rudimentary but ingenious apparatus cobbled together from whatever was available. The war changed this. To tame nuclear fission and exploit microwaves, achieving challenging objectives against strict military deadlines, demanded scientific collaboration and resources on an unprecedented scale. The prototype for this new style of research centre was Los Alamos, a self-contained laboratory town built from scratch near Santa Fe in New Mexico under the supervision of the US Army Corps of Engineers. It had all the facilities needed for thousands of atomic bomb scientists, support staff, and their families. Other vast installations were built at Oak Ridge, Tennessee, and Hanford, Washington, to supply fissile material. Another huge wartime project was the Radiation Laboratory at the Massachusetts Institute of Technology for the development of radar and microwave techniques. As well as being big, these projects also had an international aspect, with the United States sharing ideas and manpower with its wartime allies. Among those working at these laboratories were Niels Bohr, spirited out of occupied Denmark, and many European scientific emigrants – Enrico Fermi, Hans Bethe, Rudolf Peierls, Victor Weisskopf, Otto Frisch, John von Neumann, Edward Teller, Eugene Wigner, ...

After the war, their objective accomplished, these scientists wanted to forget the trauma of developing weapons of mass destruction against the clock. Remembering how Albert Einstein had been influential in 1939 in convincing President Roosevelt of the need to invest in nuclear weapons research, Los Alamos Director J. Robert

Oppenheimer thought it an opportune time to start a migration away from weapons, again with Einstein's endorsement, but this time under the banner of the new United Nations. The UN emerged at the end of the Second World War as the custodian of a hard-won but fragile peace. To extend its reach, it established specialist agencies, including the UN Educational, Scientific and Cultural Organization (UNESCO) and the World Health Organization (WHO). A UN agency laboratory would provide an ideal focus for wartime physicists to return to their peacetime intellectual pursuits.

But with the UN still in an embryonic stage, it would take some time before it could become a patron of science. In the meantime, the United States itself had to fulfil that role. Physics had changed the course of the war, and the US government had seen its strategic power. To maintain the initiative, a continual supply of scientific manpower had to be assured. One who had worked at the wartime MIT Radiation Laboratory was Isidor Rabi, who won the Nobel Physics Prize in 1944 for his pioneer work in developing the techniques that led to nuclear magnetic resonance, now a basis of imaging for medical diagnostics. Rabi understood how the big wartime laboratories functioned. With Norman Ramsey, who had worked at Los Alamos, he led the creation of a major new US scientific centre on Long Island, New York, alongside what had been a military transit camp in two world wars. This new Brookhaven National Laboratory was run as a joint venture by a partnership of major US universities. To do its work, it would be equipped with some of the largest physics installations in the world, including nuclear reactors and huge new particle accelerators. The world of physics had changed since Ernest Rutherford had probed the nucleus by random sniper fire using natural alpha particles. To reveal its innermost structure, the nucleus had instead to be bombarded by heavy artillery, concentrated beams of particles accelerated to high energies by radio-frequency pulses as they whirled round and round in orbits controlled by powerful magnets. Brookhaven's 'Cosmotron', at the time the most powerful such accelerator in the world and the first of many more cathedrals of science, came into operation in 1953.

Meanwhile, the family of the United Nations widened as traditional colonial ties fell away in the post-war world order. The new nation of Pakistan was admitted on 30 September 1947. Overseeing initial diplomatic arrangements was the distinguished figure of Sir Muhammad Zafrullah Khan, trained as a barrister in London and who from 1935–41 had been a member of the British Governor-General's Executive Council in India, before becoming India's Agent-General in China during the early years of the Second World War. Subsequently, he served as a judge in India's Supreme Court. When Pakistan came into existence, Mohammad Ali Jinnah asked Zafrullah Khan to become the nation's first Foreign Minister. It was hard to think of anyone better qualified. However, Zafrullah Khan, born an Ahmadi Muslim, had been vilified by orthodox religious factions, even before his nomination. Jinnah nobly defended Zafrullah Khan, but died, exhausted, in September 1948, only one year after Pakistan came into being. It was a tragedy for the new nation, and Zafrullah Khan lost his closest ally.

As the Second World War neared its end, Britain, wearied and debilitated by the struggle, prepared to turn its back on its colonial past, and the United States emerged as a new world imperial power. Aware of the parallels between the fast-evaporating British influence in India and in Palestine, Zafrullah Khan had been a member of an Indian delegation to the US in August 1946. Before returning to India, he stopped over in the UK, and it was during this stay that he intercepted Salam at the dockside in Liverpool in October.

Salam had first seen Zafrullah Khan in 1933 at an Ahmadi gathering, when the lawyer was a member of the Punjab Legislative Council. The upright, distinguished figure had impressed the eight-year-old boy. Later, in 1940, Salam's father had written to Zafrullah Khan, knowing well by then that his son had talent, seeking advice. Zafrullah Khan's counsel was threefold: that the boy should look after his health; secondly that school lessons should always be well prepared for, and immediately revised afterwards; and thirdly that all journeys should be used to broaden the mind as well as to fulfil their immediate objective. In addition, Zafrullah Khan pledged to pray for the young Abdus Salam. In 1946, Zafrullah Khan had unexpectedly been on the quayside when Salam's boat docked at Liverpool. The young Salam, overburdened with baggage, inadequately dressed for the British climate, and generally unsure about what to do, had been very grateful for Zafrullah Khan's help. The stern and authoritarian politician, seeing Salam's predicament, had been happy to give it.

Zafrullah Khan continued to serve as Pakistan's Foreign Minister until 1954, a fitting responsibility for a man who thought that travel broadened the mind. The next meeting of Salam and Zafrullah Khan came in 1951, while Salam was working at the Institute for Advanced

Study, Princeton, when Zafrullah Khan had to visit the United Nations. The two toured New England together. Later Salam visited the United Nations in New York, where Zafrullah Khan had pleaded eloquently for the cause of other new Islamic nations as they emerged from their colonial past, as Pakistan had done. Speaking of a visit in 1955, Salam wrote later 'This was my introduction to the United Nations. I remember entering that Holy Edifice in New York ..... and falling in love with all that the organization represented – the Family of Man in all its hues, its diversity, brought together for Peace and Betterment.'[1] With hindsight, he added 'I did not then realize how weak an organization it was, how fragile and frustrating its inaction'[2], but the spell had been cast. It was Zafrullah Khan who introduced Salam to the world of international diplomacy and politics.

In the 1953 anti-Ahmadi riots in Pakistan, Zafrullah Khan and all other high-ranking Ahmadis became highly visible targets. After moving to the International Court of Justice in the Hague, from 1961–4 Zafrullah Khan was Pakistan's permanent representative at the UN, serving as President of the UN General Assembly in 1963–4. After his stint at the UN, Zafrullah Khan returned to the International Court of Justice, and was its president from 1970–73. After retiring from the Court, like many Ahmadis he lived in exile in London, where he frequently came to Salam's home for Sunday breakfast, this being the only time he could spare from his dedicated work at the London Mosque, translating holy books into English. His legacy includes an authoritative English version of the Holy Qur'an. Zafrullah Khan was a man of deep religious principles, a scholar and a consummate politician, but living alone he had banal problems with his laundry and the upkeep of his clothing, which Mrs Salam took care of. Several years later, Salam visited Zafrullah Khan in hospital, laid up with spinal problems. As a gift, he brought *Shamail-i-Tirmizi*, a book about the personal life of the Prophet Muhammad, adding that one day, when he had leisure, he would like to translate it into English. A few months later, Salam went to see Zafrullah Khan again, this time at home, and was amazed to receive a printed translation of the book, with a dedication 'with deep gratitude to Abdus Salam, eminent physicist, with whom the idea of this book originated'. In 1980, Salam travelled with Zafrullah Khan, then aged 87, to Morocco, for the inaugural meeting of the country's Academy, where the two men were honoured, Salam for his contributions to science, and Zafrullah Khan for his influential role at the UN in the lead-up to Morocco's independence in 1956.[3] Zafrullah died in Pakistan in 1985.

While Salam had to hide during anti-Ahmadi pogroms in Pakistan, a new UN programme was gathering momentum. The 'Atoms for Peace' movement was dramatically launched by US President Dwight D. Eisenhower's December 1953 address to the UN General Assembly advocating stricter arms-control measures, and the establishment under UN patronage of what would soon become the International Atomic Energy Agency (IAEA – a body now customarily referred to as the world's 'nuclear watchdog') with its headquarters in Vienna. Another direct result of Eisenhower's initiative was a UN Conference on the Peaceful Uses of Atomic Energy, which took place in Geneva, Switzerland, in August 1955. This was one of the most influential scientific meetings ever held, presided over by Homi Bhabha, the spiritual father of India's nuclear programme. A total of some 1500 delegates from 73 nations participated. A centrepiece of the event was a swimming-pool-type research reactor flown out from the US and assembled during the conference. President Eisenhower took time out from the concurrent Geneva summit conference to view the reactor, which subsequently stayed in Switzerland. Despite the glacial climate of the Cold War, the mood was nevertheless optimistic. Scientists from all over the world were talking with each other: huge volumes of secret information were declassified and made generally available. Salam was appointed as one of the twenty scientific secretaries of the conference, responsible for keeping track of a portion of the scientific programme and for documenting the contributions. The meeting, attended by many pioneer figures of nuclear physics, was another introduction for Salam to the world of international relations.

In 1958, Salam, now at Imperial College, London, returned to Geneva for the Second UN Atoms for Peace conference, even larger than the 1955 meeting. Secretary General of the 1958 Conference was Sigvard Eklund, soon to become the IAEA's second Director General and a close collaborator in Salam's schemes. Holding these big scientific meetings in Europe was symbolic. The Old Continent had suffered badly in the Second World War, and large parts of it still lay in ruins. Manpower was in short supply and economies, anaemic after six years of war effort, had not recovered. Nobody could speak for the whole continent, but there was a collective feeling of guilt at having caused so much strife and inflicting it on the rest of the planet. Influential leaders saw the need for a united European front, and from these seeds grew what became the European Community. Science in Europe had suffered too. As well as material damage, talent had been lost in a mass 'brain drain' emigration to the United States. In the 1920s and 1930s, eager young US students had come to Cambridge, Copenhagen or Göttingen to learn about quantum theory and subatomic physics. But no more. The pendulum of science had swung across the Atlantic. To help bring it back, European visionaries called for international laboratories, administered along UN lines, where young researchers could work without having to emigrate to the US.

In 1950, Isidor Rabi was one of the US delegates to a UNESCO general meeting in Florence where these ideas were aired. After his key wartime role and now a Nobel prize, Rabi became prominent in the administration of US science. Rabi's plan at the Florence meeting was to present the US Brookhaven National Laboratory idea as a template, and substitute entire European nations for the collaborating universities that oversaw the New York centre. The Europeans had already seen the plans for the Cosmotron at Brookhaven and knew that US scientists were planning even larger, more powerful, machines. The Europeans also wanted to aim high. That UNESCO meeting in Florence set in motion a train of events that led to the creation of the European Organization for Nuclear Research, usually known as CERN, the French acronym for Conseil Européen pour la Recherche Nucléaire, established in Geneva, Switzerland, in 1954 as a joint venture of Western European nations.

These new international developments fertilized a seed that had been planted in Salam's mind during his three years as a Professor at Government College, Lahore. The Punjab may be torridly hot in summer, but the research temperature at Government College was glacial. Salam was a researcher and an inspiration to research, not a teacher of fundamentals. Research needed an infrastructure that in Lahore had been totally absent. At Cambridge and at Princeton's Institute for Advanced Studies, Salam had profited from continual visits of distinguished scientists, from seminars on the latest developments, from draft papers sent by airmail, and from comprehensive well-stocked libraries that received journals by airmail. But Lahore had no visiting scientists; no physics research was done there; its library carried out-of-date journals. Salam eagerly looked forward to summer visits to Cambridge or Birmingham to work with Paul Matthews, but this only heightened the anguish of isolation on return.

Moving from Lahore to Cambridge in 1954, Salam frenziedly turned to research once more. But as his research pulse beat faster, he remembered the academic loneliness and stultification of Lahore. Research needs intellectually fertile soil. He poignantly described it by quoting a fifteenth-century astronomer, Saif-ud-din-Salman, who had left his home to live and work in the famous observatory of Ulugh Beg at Samarkand. He had written to his father 'Admonish me not, beloved father, for asking you thus in your old age and sojourning here in Samarkand.... I love my native Kandhar and its tree-lined avenues and I pine to return. But forgive me, my exalted father, for my passion for knowledge. In Kandhar there are no libraries, no quadrants, no astrolabes, My star-gazing excites nothing but ridicule and scorn. My countrymen care more for the glitter of the sword than for the quill of the scholar. In my own town, I am a sad pathetic misfit.'[4] Salam had seen that science cannot grow flourish in an intellectual desert. 'If Einstein had been born in Burkina Faso, he would never have become what he was'[5], he pointed out.

Moving in new circles at the Atoms for Peace meetings in 1955 and 1958 underlined the possibilities of the new international order. The United Nations could provide a framework for a new scientific venture. The seed planted in Salam's mind in Lahore had now been fertilized, but still lacked nutrient. It came from a strange cycle of serendipity. Salam had left his home country in 1954 to escape the trauma of the anti-Ahmadi riots, but his departure from Pakistan and subsequent career went largely unnoticed until a major article appeared in the *Pakistan Times* of 25 August 1957, written by Mian Iftikharuddin, the paper's founder and director, and an Ahmadi. It highlighted how Salam, still only 31 but now a professor at one of Britain's leading universities, and famous after a series of research breakthroughs, was on the crest of a wave. Pakistan suddenly remembered its exile, and in December he returned to receive an honorary doctorate at the University of the Punjab, Lahore, the first of some 40 degrees *honoris causae* in his lifetime. Soon afterwards he was honoured with a national medal and a cash award of 20 000 rupees by Pakistan's President, Iskander Mirza.

In Pakistan, Salam met the nation's new strongman, General Ayub Khan. The first Commander-in-Chief of Pakistan's Army, he was soon to replace Iskander Mirza as President in September 1958 after a bloodless coup that effectively suppressed anti-Ahmadi demonstrations. Ayub's objectives were wider than simply gaining personal power, and his enlightened thinking introduced an autocratic political structure, together with rigid measures to stabilize Pakistan's economy. Under the blanket of his martial law, a key component in Ayub's perception of the nation's future was a new scientific and technological thrust. Cut off from the rest of the subcontinent, science had stalled in Pakistan. Salimuzzaman Siddiqi, a Muslim, and head of the Indian Council for Scientific and Industrial Research, had visited Pakistan in 1948 and reported to Prime Minister Nehru that 'neither science nor arts exist'[6]. Pakistan Prime Minister Liaqat Khan requested Nehru to allow Siddiqi to move to Pakistan, where he went on to head the corresponding national council, and initiated programmes in chemistry and botany. It was the first step.

Ayub Khan's fresh scientific and technological plan for Pakistan was unveiled at the 1958 meeting of the Pakistan Association for the Advancement of Science, loosely modelled on the successful British Association for the Advancement of Science. The 1958 meeting was carefully stage-managed, opened by the impressive figure of President Field Marshal Ayub Khan, with Prince Philip, Duke of Edinburgh, as guest of honour. Speaking on the role of scientific co-operation in the English-speaking Commonwealth, Prince Philip referred to Salam, sitting in the audience, as an epitome of scientific endeavour[7]. Salam felt on firm ground. He had heard what his Imperial College patron Patrick Blackett had said on the subject of 'Technology and World Advancement' at the meeting of the British Association for the Advancement of Science in Dublin in 1957. Technology, Blackett had said, is not limited to mere technical know-how, but in a wider context extends to its influence and impact on all spheres of modern life. This impact had been considerable in the West, and Ayub Khan's plan was to introduce modern science and technology to Pakistan. Now he saw who could help achieve this. Salam was appointed to Ayub Khan's Scientific Commission in 1959, charged with making recommendations for new directions in Pakistan's science effort. He later became a member of Pakistan's newly established Atomic Energy Commission, qualifying him to become a delegate at the International Atomic Energy Authority in Vienna. In rapid succession, Salam also became an advisor to the national education commission and finally the President's Chief Scientific Advisor. It was a post he was to hold for 13 years, the apogee of his career in Pakistan. At President Ayub's side, Salam finally became the civil servant that his father had always wanted him to be.

For his new role, Salam, working in Europe, needed close collaborators in Pakistan. Salam met Ishrat Hussain Usmani by chance on a railway train during a visit to Pakistan in 1957. Usmani had earned a physics PhD at London under Nobel Laureate G.P. Thomson in the 1930s and moved to the Indian Civil Service, later opting for Pakistan. When Salam encountered him, he was running the Pakistan Geological Survey, having previously served as Director General of Customs, Imports and Exports. The Pakistan Atomic Energy Commission had been created in 1956, but with no knowledgeable nuclear physicists in the country was moribund until, at Salam's urging, Usmani took over the helm in 1960. To breathe life into the organization, students were sent to study abroad in the early 1960s. As a core activity, Salam and Usmani launched the Pakistan Institute of Nuclear Science and Technology (Pinstech). The striking Islamabad centre was designed by architect Edward Durrell and its reactor went critical in December 1965, going on to play a major role in producing fuel for Pakistan's first nuclear power plant, which began to supply power to Karachi in 1972.[8] At this stage, Pakistan's nuclear efforts were entirely dedicated to peaceful applications. As well as the reactor programme, research centres were set up for the use of radiation and isotopes in agriculture, nuclear medicine and radiotherapy. Salam's proposed overhaul of Pakistan's science and technology in this era could have been modelled on the modernization of Japanese education in the nineteenth-century Meiji restoration, or the wide-reaching reforms of Peter the Great as Emperor of Russia (1682–1725).

Other areas of technology were also boosted in the Ayub era. Water has a special importance in Pakistan. The flat delta of East Pakistan (which in 1971 became the new nation of Bangladesh) has too much of it and is frequently flooded. On the other extreme, the Punjab would be desert if it were not for mighty irrigation schemes. Its plight was amplified after the partition of British India, when the new republic of India retained control over a large part of the headwaters of the Indus valley. To avoid a fresh face-off with India, in 1959 Ayub Khan negotiated a deal with the World Bank, including a substantial Indian contribution, for new dams, barrages and canals to safeguard the Punjab water supply. Under the Indus Waters Treaty, signed in 1960, the eastern Punjab rivers, the Ravi, Beas and Sutlej, would be available to India, while to the west, the Jhelum, the Chenab, and the Indus itself would be Pakistani resources.

But in the rural farmland there was another problem: continual seepage from the artificial canals constructed by the British had waterlogged surrounding farmland. At the same time, the hot sun had boiled away so much water that the soil had become encrusted with leached salts. To ensure the agricultural health of Pakistan's breadbasket province, Salam realized that drastic measures were needed. The 1961 meeting marking the centennial of the Massachusetts Institute of Technology included a session on Science for Developing Countries. Salam heard Patrick Blackett – his boss at Imperial College – proclaim that all the technology that underdeveloped countries needed already existed somewhere else in 'the world supermarket of science'. All that was required was to find it and buy it off the shelf. Salam objected, saying that skilled manpower was first needed to read the labels on the supermarket shelves. He then gave the specific example of the Punjab marsh, which had not been studied anywhere. The mathematician Norbert Wiener, then scientific advisor to President Kennedy, heard the plea and took Salam aside.

Wiener knew that Kennedy would soon be welcoming Pakistan President Ayub Khan on an official visit to Washington and wanted to be able to offer Pakistan something other than arms. The problem of the Punjab marsh was drawn to the attention of the US President, and when Ayub Khan went to Washington he learned that the United States was eager to help. A team under eminent oceanographer Roger Revelle was given the brief[9]. Artesian wells had already been sunk to boost drainage, but far too few of them. Revelle and Salam's solution was to sink many more wells and pump over as wide an area as possible, so that more saline water would be removed than would seep in from the surroundings. Although it was not his idea, Salam was proud that the underlying reasoning – that area increases faster than periphery – had also been used by Patrick Blackett in the British Naval Operational Research unit in the Second World War. The apparent vulnerability of a large merchant convoy to submarine attack was more than compensated by the increased number of surrounding naval escort vessels. A convoy of sixty merchant ships with twelve escorts was better protected than two convoys of thirty ships, each with six escorts.

On arrival in Pakistan, Revelle realized that saline waterlogging was not the only problem in Punjab agriculture, and recommended a modernization programme to educate farmers and improve roads, the supply of fertilizers and pesticides, grain storage, etc. The result was a green revolution that changed the face of Pakistani agriculture, transforming the Punjab into Pakistan's breadbasket, with grain and crops for export. The only regret for Salam was that the ideas had to come from American, rather than Pakistani, brains.

Another Salam contribution came with the expansion of Pakistan's university system. When the nation was created in 1947, its capital was the great port city of Karachi. However, this led to an overconcentration of resources that handicapped the rest of the country. In 1958, during Ayub Khan's administration, a 'green-field' site immediately north of Rawalpindi was chosen for a new permanent capital, with planning and construction largely headed by Greek architect Constantinos A. Doxiadis. The impressive new city, with much greenery and open space, needed its university. Salam's international skills helped win new funding, and he emphasized the importance of establishing a strong physics group at Islamabad's new university, later renamed Qaid-i-Azam University by Ali Bhutto in honour of Pakistan's founding father, Muhammad Ali Jinnah. With the older universities restricting themselves to the type of teaching institution that Salam had encountered in Lahore, Islamabad was the first with the aim of doing research, and providing research training. It was rapidly populated by young Pakistani scientists who had trained in the West, many of them in Salam's group at Imperial College, which for a long time included a noticeable Pakistani profile – in the mid-1960s the 30-strong group included half a dozen of Salam's compatriots. Riazuddin, his student from Lahore days, founded the theoretical physics group at Islamabad, whose founder members also included Riaz' twin Fayyazuddin, and Faheem Hussain, both Imperial College PhDs.

Ayub Khan was a great supporter of Salam's vision for Pakistan, and Salam's writings reflect his own pride in his work for his country, helping a new nation belaboured by difficulties to get on its feet. It is easy to get the impression that Salam was Ayub Khan's right-hand man. However, Ayub Khan's political autobiography *Friends not masters*[10], covering the period up to 1965, does not mention Salam. In the swathe of Reform Commissions set up in the wake of Ayub Khan's proclamation as President, that on science was ninth, after land, maritime affairs, law, administrative reorganization, education, the location of a federal capital, credit, and food and agriculture.

After Ayub Khan's rapprochement with China in the early 1960s, in March 1965, Salam accompanied the Pakistani President on an official visit to Beijing. There, he inspected Chinese research reactors and lectured to the Chinese Academy of Sciences on recent researches in subnuclear physics. At that time, there were probably only a handful of people in China who understood such matters, and Salam's talk in front of 600 people was probably largely incomprehensible. During his trip, Salam was impressed by Premier Chou En-Lai's detailed knowledge of China's scientific effort, and by the nation's ambition and pragmatism in attacking major projects, where detailed planning and foresight avoided many last-minute problems. Salam attributed this to the Chinese 'intense involvement with their own history', adding 'History seems to live with the Chinese people in a much more meaningful way than it does for us'.[11]

In August of that year he attended a meeting of Pakistan's premier scientists in the mountains of Swat. Ayub Khan was there, but had many things on his mind. Salam proposed a new order of priority in national science matters, with its role in defence prominent, followed by university science, medicine and public health, food and agriculture, irrigation, and industry. The stern figure of Ayub Khan listened politely. On arrival in London after the meeting, Salam was shocked and humiliated to learn that Pakistani troops had invaded Kashmir and war had broken out again between Pakistan and India.

These new responsibilities in Pakistan meant extra travel, adding to his workload. Diplomatic trips required being briefed beforehand and extensive background research, which he always did himself. However, physics research was still his first priority. One of Salam's great pleasures was attending international physics meetings, meeting old friends and colleagues, and hearing the latest physics news, most of which he already knew, but there was always a new slant, a fresh opinion, an unexpected result, a research rumour. One such meeting came in June 1960 at the *castelletto* adjacent to the eclectic Miramare Castle, on the Adriatic Coast, seven kilometres north of the city of Trieste in Italy. The ornate mid-nineteenth century mock fortress had been built by the Habsburg Archduke Maximilian, brother of the Emperor Franz Joseph and grand admiral of the Austro-Hungarian fleet, for his beloved queen, Charlotte, daughter of the King of Belgium. Maximilian's career had taken an unexpected turn in 1864, when Napoleon III installed him as Emperor of Mexico in a grandiose plan to extend the French Empire while the United States, torn by civil war, had its back turned. When French troops withdrew from Mexico a few years later, the unprotected

Maximilian was captured, tried and shot. Maximilian and his wife had lived in the smaller *castelletto* while the main castle was being built. The castle, surrounded by Maximilian's original landscaped gardens, is a poignant monument to a lost empire and a tragic romance. It was a fairy-tale venue. About 30 scientists had been invited to a small informal meeting on the physics of elementary particles, organized by the local scientist Paolo Budinich. It was the first meeting between Salam and Budinich, and Salam's first encounter with Trieste. The meeting and its setting enchanted him.

Soon after, in September 1960, Salam attended that year's international meeting of particle physics at Rochester University, New York. These 'Rochester' meetings were the focus of the subnuclear physics calendar. In addition to the latest scientific results and gossip, physicists would exchange ideas for new laboratories and international consortia. In these discussions, the idea of a UN-based laboratory, first mooted by Oppenheimer in 1945, periodically resurfaced. At the 1960 Rochester meeting, the after-dinner speech was given by John McCone, then Chairman of the US Atomic Energy Commission, who looked at the possibility of creating world-wide physics research centres, principally to build the next generation of particle accelerators. These future machines for the world would take over from those currently planned in regional initiatives in the US, Europe and the USSR.

After the dinner, Salam and his former teacher, Nicholas Kemmer, were drinking coffee with Hans Bethe, who Salam had first met at Princeton and later at Cambridge, and Robert Sachs, one of the first generation of US theorists to earn a research degree under Oppenheimer, and who Salam was later to meet frequently at summer meetings in Madison, Wisconsin. All four were theorists, and immediately gave McCone's idea a theoretical spin. Instead of an international centre for a large machine, with all its problems of vast funding and complicated civil engineering, why not have instead a smaller, theoretical centre, something closer to the original Oppenheimer plan, where scientists from different countries could meet and work together? The vast new laboratories in the US required annual operating budgets of hundreds of millions of dollars, even when their civil engineering and construction work for their huge machines was complete. CERN in Geneva was funded by a consortium of European nations who contributed to a similarly large annual budget in proportion to their gross national product. An international theoretical centre would require only a tiny fraction of such an outlay. At the time, collaboration, not to mention just contact, between East and West was difficult because of the Cold War. An international theoretical centre on neutral territory could sidestep such political barriers. It would also overcome the problem that had long nagged at the back of Salam's mind, the isolation of scientists working in poorer countries. With modest international funding, these researchers could transfer periodically to such a centre to renew their contacts, recharge their intellectual batteries and generate new research momentum.

Just several weeks after he had heard McCone's speech at the Rochester physics meeting, Salam was Pakistan's delegate at a General Conference of the new International Atomic Energy Agency in Vienna[12]. Here, he tentatively launched his idea for an international theoretical centre. He pointed to the role of theoretical physicists – Bohr, Einstein and Fermi – in the emergence of man's use of nuclear energy. 'The time has come when the Agency might pay back the debt by considering if it might sponsor an international institute for theoretical physics.' But fine speeches were only one weapon in his diplomatic armoury: schooled by Zafrullah Khan, with experience from two major international meetings in Geneva, and with the confidence of Ayub Khan's backing, Salam set about his personal political business in Vienna. He was not just an envoy simply relaying his government's wishes. He was also his own ambassador. Brought up in a crowded single-roomed house in a town that did not even have electricity, he was now negotiating at a world level. Using his intellect as a weapon and the predicament of the Third World as a flag, Salam's new mission was to battle against the cruel imbalance of global wealth and resources. The injustice of the world angered him, but this was no hot fury that could be vented in rage. It was instead a cold anguish, a deep anger against the accumulated inequity of mankind and that sublimated in Salam as a relentless drive. The mission would continue for the rest of his life.

Salam did not yet know if the infant IAEA would be interested in a project outside its 'nuclear watchdog' brief. An international theoretical centre for subnuclear physics was a good thing, but it could detract attention from the IAEA's baseline responsibility for safeguarding nuclear investment and controlling the spread of nuclear weaponry. In an earlier exploratory trip to Vienna, he had gone to see Munir Ahmad Khan, who had been his contemporary at Lahore's Government College from 1942–6. Like Salam, Khan had gone overseas to continue his studies, first a masters in electrical engineering from North Carolina State University, followed by postgraduate work in nuclear engineering at the Illinois Institute of Technology. Also like Salam, he had been a scientific secretary of the UN Atoms for Peace meeting in Geneva. In 1958, he became the new IAEA's first staff member from a developing country. Salam brought the germs of his new idea to his compatriot's office in Vienna, who introduced him to some IAEA notables.

Khan relates how Salam returned to Vienna in 1960 to formally plead his case before the IAEA. 'He landed at the airport dressed in his top-coat and a canvas hat which was not only discoloured but rumpled out of shape.' (Salam was invariably overdressed. Because of his upbringing and the rigours of the European climate, he wore headgear outdoors, but this was not always as impressive as the turban he sported at the 1979 Nobel Prize Award.) 'I was horrified to see him in that attire,' continued Khan. 'I persuaded him to take his coat and hat off, which I deposited in the trunk of the car. Between the airport and the hotel, I managed to make the hat disappear. When he reached his hotel room he started looking for it. I reminded him that he had to rush to the Board meeting and the search for the hat could be resumed later.'[13] The hatless Salam's resolution was co-sponsored by Afghanistan, the Federal Republic of Germany, Iran, Iraq, Japan, Pakistan, the Philippines, Portugal, Thailand and Turkey[14]. It was unanimously approved, but with eleven abstentions. Among the abstaining delegates was that of Australia, with the disdainful comment 'theoretical physics is the Rolls-Royce of sciences – the developing countries need only bullock carts.' However, the idea was on its way, and Salam moved up a gear and began networking the world's scientific élite. As sometimes happens, the press in its haste grasped the wrong end of the stick: the *Pakistan Times* reported that Abdus Salam was to become the director of a new nuclear authority in Vienna: questioned, an obviously puzzled Salam was 'unaware' of such an appointment.

Under the chairmanship of IAEA Research and Laboratories Director Carlo Salvetti, an IAEA study group was commissioned. Its members were all theoretical physicists, and therefore acquainted with Salam's work: two of them – Nicholas Kemmer and Jacques Prentki – had actually worked with Salam. Significantly, the group included a UNESCO representative. Paolo Budinich from Trieste was an official observer. Three members of the group came from Denmark, which was keen to provide a home for such a centre. The figure of Niels Bohr in Copenhagen had towered over European physics for half a century. Apart from short periods during two World Wars, in the First as a student at Manchester under Rutherford, and in the Second at the Los Alamos atomic bomb project, he spent his entire career in the Danish capital, which had become a focus for theoretical physicists from all over the world. Post-1945, Bohr worked energetically for world peace, and had promoted the 1955 Atoms for Peace conference in Geneva. He was also one of the major proponents of the European CERN laboratory in Geneva, and his Copenhagen centre had been the provisional home of CERN's infant Theory Division while its new site in Geneva was being readied. (Bohr died in 1962, while plans for Salam's idea of an international theoretical centre were still in the melting pot.)

The study group unanimously supported the plan, and recommended that it be truly international, looking beyond the UN-IAEA community, so that, for example, the People's Republic of China (as yet not a UN member) could become involved. It could act as a role model for future such ventures in other fields of study. The operations of other UN agencies, such as the IAEA or the World Health Organization, were oriented to field work by specialists working all over the globe, with their headquarters mainly as administrative and support centres. On the other hand, a centre for theoretical physics would be a place where specialists from all over the world would converge, providing a critical mass for them to do their work. The study group also listed logistical criteria that the future home of the institute should satisfy. All of them fitted Copenhagen, a capital city with good communications, a tradition of physics research and a strong university tradition. However, Salam could also mention other possible sites – Geneva, Vienna, Dubrovnik, Stockholm and Warsaw. Italy had now mounted a strong counteroffer, with three possible sites – Trieste, Florence and Naples. Faced with such profusion, the study group recommended that the eventual site should be acceptable to West and East.

The study group's findings went to the IAEA's Scientific Advisory Committee, who demurred, suggesting that the same, or at least similar, results could be obtained without setting up a new centre. Instead, fellowships could be set up at existing institutions, and supplemented with regular summer schools, once university coursework was over for the year, in different venues. The committee suggested that a few such schools could be organized to test the idea. Such countersuggestions had emerged before, usually from prominent scientists in major countries. The proposal and all the ensuing findings and recommendations came together in a crowded agenda at the next IAEA General Conference in the summer of 1961. There was a lot for the delegates to get through. With Sterling Cole stepping down, the meeting had to elect a new Director-General, always a difficult task in an international setting, with each country or bloc pushing for its candidate, and the infant organization also had much other important administrative and legislative business to get through. Sigvard Eklund from Sweden was elected as the new Director General. The Soviets had been pushing for their own candidate and were initially outraged by the decision, storming out of the room and threatening to break off relations. It is a measure of Eklund's diplomatic skill that he was subsequently reappointed four times with the full support of delegates.

Although the two men came from very different backgrounds, Sigvard Eklund was to play a major role in Salam's career. Born in 1911 in Kiruna, Sweden, from 1937–45 Eklund worked at the Nobel Institute of Physics under Manne Siegbahn, eventually becoming Senior Scientist, before taking up a similar post at Stockholm's Research Institute for National Defence. In parallel, he was Assistant Professor of Nuclear Physics at the Royal Institute of Technology in Stockholm. From 1950, he moved steadily upwards at the Swedish Atomic Energy Company, AB Atomenergi: first Director of Research; then Deputy to the Managing Director; and Director of the Reactor Development Division. He and Salam first knew each other from the Geneva Atoms for Peace conferences.

Directors General of major international organizations have their attention pulled in many directions at once, but Eklund soon got around to sending IAEA Member States a questionnaire about the proposal for a theoretical physics centre and his Scientific Advisory Committee's findings. Despite its distance from the IAEA's core business, Salam's scheme resonated with Eklund, and he worked tirelessly towards its success. Few countries replied to his initial request, but those that did ranged from enthusiastic (including Pakistan) to disinterested and even dissenting. Meanwhile, Denmark continued to push hard for a new centre grafted onto what already existed in Copenhagen. The Italian proposal, now focused on Trieste, steadily gathered momentum. Other offers of a home came in from Pakistan (Lahore) and from Turkey (Ankara), fitting in well with a broader international outlook, but neither could claim a strong existing university base and could not compete seriously with any of the European offers.

A school/seminar in the summer of 1962 was the trial run of the IAEA Scientific Advisory Committee's alternative scheme – fellowships at existing universities supplemented by annual summer schools. The trial school was held in Trieste, and at the school the idea of a new centre was as much a talking point as the physics. Salam and Budinich urged participants, once they returned home, to write letters of support for the scheme and to canvass their IAEA delegates. The other arm of the Scientific Advisory Committee's counterproposal, the establishment of IAEA fellowships at major research centres, did not look promising. Of the four research centres solicited, two – Copenhagen and CERN – did not welcome the idea at all, while two others – the Joint Institute for Nuclear Research at Dubna, near Moscow, and the Institute for Advanced Study, Princeton – were only mildly encouraging.

The results of all this came together in the September 1962 General Conference of the IAEA in Vienna's elegant Neue Hofburg. The Scientific Advisory Committee still advocated substituting Salam's idea by fellowships and regular schools. Many major national delegations were hostile to Salam's scheme. The US, represented by Harry Smyth, said the time was not yet ripe, as did Belgium, Canada, France, the Netherlands and the UK. After the unusual spectacle of the USSR agreeing with US opinion, Czechoslovakia, Hungary, Poland and Romania lined up behind the Soviets. The German delegate had voted in favour of an earlier resolution, and had been reprimanded afterwards by his government. But the meeting in Trieste had produced a shower of sparks, some of which were still smouldering. The outcome hung in the balance. Salam's ten-minute speech on 22 September was one of the most eloquent of his life. First, he outlined the objections to the establishment of an international research centre for theoretical physics:

1. Does such research fall within the scope of the Agency's objectives?
2. Do physicists from emerging countries really need and desire such a centre?
3. Can such a centre be created and can the Agency afford it?

After setting up his straw men, he carefully demolishing them one by one in his distinctive husky voice whose logic nevertheless thundered and resonated. On one side, there were countries desperate for new science and technology: on the other, rich countries who already had it. The Scientific Advisory Committee's fellowship alternative did not go far enough. 'What is needed at this stage,' he exhorted, 'is an active international centre, sponsored by an international body like this Agency. Only then can first-rate men from less privileged countries come periodically as of right to relive with their peers – the pioneers and thinkers of the international world – and thus give of their best.' Salam concluded 'Let us project ourselves twenty years from now. The world is moving closer, economically, intellectual, scientifically. In twenty years, there will be international research centres not only for theoretical physics but for most fundamental sciences. The world trend is in this direction and nothing can stop it. It is possible for us in this agency to take the initiative in forwarding this movement....I commend to you the resolution in front of us'.[15] Afterwards Salam admitted that during that meeting he had smoked some fifty cigarettes (he seldom smoked) and consumed about a kilo of grapes to sustain his blood sugar level.

The speech fanned those smouldering sparks from the Trieste meeting into a blaze, producing a torrent of votes from the have-not nations that flooded over the barrier erected by the major powers. It was a Third World David versus an unusual Goliath, provided by the Western and Eastern bloc powers voting in unison. Rabi, who had served as chairman of the US Atomic Energy Commission in the wake of Oppenheimer's resignation amid allegations of disloyalty, and was now a member of the IAEA Scientific Advisory Committee, admired Salam's virtuoso performance. With no votes against and only four abstentions, the proposal, which had looked in peril of foundering, was carried and moved forward into a feasibility study. The outcome showed how well Salam had done his political homework, patrolling the corridors of the Vienna meetings, goading, urging. Peering through the fog of residual prejudice from the imperial era, he had seen a new direction: as Pakistan's President Mohammad Ayub Khan said in his 1967 autobiography: 'People in developing countries seek assistance, but on the basis of mutual respect: they want friends, not masters'[16].

Salam was revealing a new side of his personality. He was becoming a man of power. His objectives were so clear and his demands so insistent, that it was difficult to ignore them. The British physicist John Ziman, who had been a lecturer with him at Cambridge in the early

1950s and was later to be a close collaborator, described Salam's highly developed power of persuasion. The idea was to count on friends and colleagues that he had known for some time. There was no bullying or threats. Ziman said 'He would take you by the arm and say... "I want you to go to Valparaiso tomorrow, on mission."…. I discovered that for that request from Abdus Salam there were only three answers. One was "Well, it's against my religion to go to Valparaiso." You had to have a really strong view. He being a deeply religious and sincere man, that would have been enough. The other answer might have been "I'm very sorry but that day I must be in Singapore." The third answer was "Yes, I'll do it. What do we do? How do we start? What's to be done?"'[17] The astute Salam had been taught by Zafrullah to do his political homework well and know exactly what needed to be done. He chose his collaborators wisely and was very insistent. That was his personal formula for power. He also developed an impressive style of 'management by flattery', which worked well in the Italian environment of Trieste.

Despite the show of enthusiasm at the General Conference, the IAEA's Scientific Advisory Committee continued to be reluctant, repeating its earlier objections and advocating the alternative route via fellowships and summer schools. In Vienna, the objective was to find a way that involved minimal outlay to the Agency. The provisional name 'International Centre for Physics' had mutated to 'International Centre for Theoretical Physics', reflecting the fact that blackboards and pencils are an inexpensive way of doing scientific research. However, the insistent Italian offer of a home for the Institute continued to gather momentum. By early 1963 there were several clearly delineated offers on the table: from Denmark, with construction costs worth $800 000 and access to existing scientific equipment; from Italy, centred on a new building in Trieste with 3000 square metres of floor space, and initial annual running costs of about $300 000; Austria offered a site in central Vienna with more than a million dollars of construction funds; a similar sum was also forthcoming from Pakistan, at the University of the Punjab in Lahore; a Turkish proposal for an institute alongside the Middle East Technical University in Ankara technically missed the IAEA deadline, but did its best to catch up. To evaluate these offers, the IAEA Board of Governors asked the Director General to set up a group of three advisors to study the offers that had come in for the centre. In the face of such generosity, the Scientific Advisory Committee's call for the alternative fellowship route was muffled, although prestigious names such as Bhabha still periodically reminded IAEA of this possibility. But such voices were increasingly fewer and less insistent.

The Chairman of the IAEA Board of Governors was now Salam's Pakistani colleague Ishrat Usmani. The three 'wise men' advisors were: the US physicist Robert Marshak, who had founded the series of international conferences on high-energy physics initially held at Rochester University, New York; Jayme Tiomno from Brazil, recently Salam's guest at Imperial College, London; and Leon Van Hove, Head of the Theory Division at CERN, Geneva. All had worked at the Institute for Advanced Study, Princeton. They concluded that the decision should be between the Copenhagen and Trieste offers. With the Italians offering more running costs, thereby minimizing the burden on the Agency, the IAEA decided to accept the Italian offer. With the proposal becoming more concrete, UNESCO, which sent a delegate to Vienna meetings, came into the open, indicating that it could contribute some $100 000 of running costs, spread over several years.

With opposition from developed countries less insistent than in the past, in June 1963 the IAEA Board of Governors approved, 'on a provisional basis', Salam's plan for an International Centre for Theoretical Physics. It would be in Trieste, Italy[18]. A new name had appeared on the international map. Salam was a citizen of a country whose boundaries had emerged from the post-war partition of British India. It is remarkable that his creation also found a home inside frontiers redrawn in the aftermath of the Second World War.

REFERENCES

1. Lai, C. H., Hassan, Z., (ed.), *Ideals and realities, selected essays of Abdus Salam*, (Singapore, World Scientific, 1984)
2. Salam A., *Ideals and realities*, Bulletin of the Atomic Scientists, September 1976, Lai, C. H. and Hassan, Z., (ed.) 1984
3. Salam A., *Homage to Chaudri Zafrullah Khan*, National Perspectives, 12, 305, 1986, reproduced Lai, C. H., and Hassan, Z., (ed.) 1984
4. Lai, C. H., Hassan, Z., (ed.) 1984
5. Lai, C. H., Hassan, Z., (ed.) 1984
6. Akhtar, M., Biography of Salimuzzamann Siddiqi in Biographical Memoirs of fellows of the Royal Society (London, Royal Society, 1996) 43, 403
7. Prince Philip, *Cooperation in science in the commonwealth*, Pakistan Journal of Science 11, 128 (1959)

8. Pakatom, Newsletter of the Pakistan Atomic Energy Commission, special issue, May 1999.
9. http://ic.media.mit.edu/projects/JBW/ARTICLES/REVELLE/REVELLE.HTM
10. Ayub Khan, M., *Friends not masters* (Oxford, OUP, 1967)
11. Salam, A., report to President of Pakistan, 28 March 1965, provided by Louise Johnson.
12. Hamende, A. M., *A guide to the early history of the Abdus Salam International Centre for Theoretical Physics*, (Trieste, Consortium for Development and Research, 2002)
13. Khan. M. A., *Lifelong friendship with Abdus Salam*, in Hamende, A. M. (ed.), *Tribute to Abdus Salam*, (Trieste, ICTP, 1999)
14. Salam, A., *Physics and the excellences of the life it brings*, in *Pions to quarks*, Brown, L., Dresden, M., Hoddeson, L., (ed.) (Cambridge, CUP, 1989
15. Lai, C. H., Hassan, Z., (ed.) 1984
16. Ayub Khan M.,, *Friends not Masters*, (Oxford, OUP, 1967)
17. Ziman J. *Delight in Abdus Salam's company*, in Hamende, A. M. (ed.), *Tribute to Abdus Salam*, (Trieste, ICTP, 1999)
18. For a complete and extensively researched account of the history of the International Centre for Theoretical Physics, Trieste, see Hamende, A. M., (2002)

    Another fine source is Alexis de Greiff Acevedo *The International Centre for Theoretical Physics 1960–1979: Ideology and practice in a United Nations Institution for Scientific Cooperation and Third World Development*, Dissertation submitted for the Degree of Doctor of Philosophy, University of London, Imperial College of Science, Technology and Medicine, December 2001, and the many references therein.

11

Trieste

On a map, the Adriatic appears as a rectangular wedge that prises Italy from the great Balkan peninsula. The map suggests that Trieste, in the north-east corner of the Adriatic, might be a mirror image of Venice in the north-west corner. Nothing could be further from the truth. While the eternal splendour of Venice reigns over its lagoon and the flat Veneto hinterland, attracting hordes of tourists from across the world, Trieste hides below bleak cliffs of Alpine rock, clinging precariously to a narrow coastal strip. But Trieste was not always eclipsed by the great maritime metropolis of the Doges and Marco Polo. Its apogee came in the eighteenth and nineteenth centuries when the Habsburg and later the Austro-Hungarian Empire spanned a wide expanse of central Europe. In principle it was possible to set sail from Vienna down the Danube, and reach the Black Sea, but such a powerful empire demanded a proud deep water port and naval base. This became Trieste's imperial role, the Shanghai of central Europe, a function that became even more important after the opening of the Suez Canal in 1869.

After the First World War, a once grand Empire evaporated in the 1919 Versailles agreement that redrew the map of central Europe. One of its creations was the kingdom of Yugoslavia, inheriting much of the east coast of the Adriatic. But under pressure from the patriotic Irredentist movement, the tiny strip around Trieste became instead part of Italy, a nation that itself had been united only half a century before. The harbour of Trieste, the northernmost major port in the Mediterranean[1], cut off from its traditional hinterland, served little purpose in a country whose long coastline was studded with commercial ports. Isolated at the eastern fringe of Italy, Trieste began to wither.

After the fall of the Fascist regime in 1943, Trieste was overrun in the German occupation of Northern Italy. An old rice factory in San Sabba, on the outskirts of the city, was converted into the only Nazi concentration and extermination camp on Italian soil. San Sabba was equipped with its own crematorium, but most prisoners – partisans and Jews from Italy, Slovenia and Croatia – were herded to the larger and more infamous camps in Germany and Poland.

With Italy weakened after the Second World War, Istria became part of the new socialist republic of Yugoslavia, and Italian refugees streamed northwards. The port of Trieste too could easily have had to hoist another new flag, offering a strategic warm deep-water outlet on the east side of the newly descended 'Iron Curtain'. The name and its association with Trieste had come to international prominence in Churchill's speech in Fulton, Missouri, in February 1946. Although he was no longer British premier, when he spoke, the world still listened: 'From Stettin in the Baltic to Trieste in the Adriatic, an "iron curtain" has descended across the continent,' he thundered.

The Western powers stepped in, and for several years the coastal strip, part Italian-speaking, part Slovene, was administered as an independent free territory. In 1954 the city was finally restored to Italy, while the surrounding uplands were absorbed into Yugoslavia. At the cold edge of Churchill's Iron Curtain, connected to the rest of Italy by only a narrow thread of coast, offering barely enough space for a road and a railway, the tiny outpost felt even more isolated. With only its unique position and fine buildings reflecting its imperial Central European past, it was a city searching for a new role.

One who felt this strongly was Paolo Budinich[2], born in 1916 on the Adriatic island of Lussino (Losinj)[3] in a family with strong maritime traditions. A few years later, the family, whose self-esteem had been undermined by an imposed nationality that continually oscillated, moved to Trieste, newly but more firmly Italian. After graduating as a physicist, Budinich was caught up in the Second World War, first as a volunteer in Italian submarines. Not wanting his parents to know what he was doing, he told them he was working in the Taranto dockyard, and left a series of letters home with a colleague, who was supposed to mail them at regular intervals. The letters got mixed up, the family became suspicious, and found out what Budinich was up to. Pleading ear problems, he became instead an observer in seaplanes. In the confusion of allegiances in 1943, as Italy was overrun by armies, he resolved to escape to North Africa, make contact with the British, and get involved in the Italian partisan movement. His seaplane made an intermediate stop on the water, where it was intercepted by a British warship. In London, he tried to convince the British of his value for the partisan campaign. His handler was John Skeaping, the British artist who had lived in Italy before the war and was now working in the Intelligence Corps and the Special Air Service. With the British unconvinced, Budinich was transferred to a POW camp in the United States.

In 1945 he returned to a stateless and disorganized Trieste, and later worked as a physicist at Padua and at the Max Planck Institute in Göttingen. Before 1945, Trieste had not even merited a university, but this was soon changed, initially with a link to the ancient university of Padua. With this achieved, Budinich (the family had changed its name to the italianate Budini during the Mussolini era but would eventually revert back to the original form) believed that university science could be used as a tendril to link Trieste more firmly to Europe. In 1953 he became a professor at the new University of Trieste and the following year Director of its Institute for Physics. From this elevated seat, he was convinced that a vibrant university close to the Iron Curtain, a 'University of Central Europe' – a loose amalgam of institutes of higher learning linking Trieste to Vienna, Graz, Prague, Ljubliana, Zagreb and Budapest, would be a good political and cultural investment.

These ideas were temporarily shelved in 1954 when Budinich went to Zurich to work with Wolfgang Pauli, but the contrast between the prestigious centres of Europe and the backwater of Trieste seemed even more marked on his return. His city had to find a role to play, complementing newly emerging scientific specialities of the central European cities. He sought UNESCO backing for a loose scientific federation of central European universities in Austria, Hungary and Yugoslavia that reflected the old Austro-Hungarian allegiances. However, with the Iron Curtain now firmly in place, this was vetoed by the heavy fist of the Soviet UNESCO delegation. But this still did not deter Budinich, who in 1960 organized the international seminar on elementary particle physics at Trieste's fairy-tale Miramare castle that introduced Salam to Trieste. If Trieste had as yet no international science to offer, Budinich figured that at least its setting might appeal to international scientists. He wanted prestigious young visitors, and knew that Salam was at CERN in Geneva that summer, working with Jacques Prentki. The modest overnight train fare was within Budinich's tight budget. Although Salam was based in London, with his Pakistani credentials he could be presented as non-European, widening the geographic reach of the meeting.

At this point, two independent visions started to come together. Salam was pushing for an international centre for theoretical physics with UN agency backing but that did not yet have a home: Budinich was striving to get Italian backing for some kind of international centre in Trieste with UN backing. (Salam's plan was in fact the sublimation of an idea that been in circulation since the end of the Second World War, and had been periodically aired in high-level international conferences.) One who helped these two separate plans to resonate was Edoardo Amaldi, the prestigious Italian physicist and former colleague of Enrico Fermi who had played a key part in establishing the international CERN laboratory in Geneva ten years earlier, and who had moved through a series of influential roles, on the international side as President of the International Union of Pure and Applied Physics (IUPAP), and in Italy as President of the Istituto Nazionale di Fisica Nucleare (INFN). In 1960, Budinich met Amaldi at the Ministry of Education in Rome after Amaldi had returned from the IAEA meeting in Vienna where Salam, as Pakistan's delegate, had launched his idea of an international centre. Salam's bid at IAEA soon acquired an offer of an Italian home.

The idea was Salam's, but the key catalyst in the Trieste outcome was Budinich, committed to persuading Italy to support the scheme. His suggestions fell on fertile ground. Aware of Trieste's proximity to the Iron Curtain, the Italian government was keen to boost the infrastructure and international importance of the Adriatic port. Budinich achieved rapid results with his wide university connections and by skilfully short-circuiting cumbersome official channels. When it came to finance, he felt himself in deeper water, and was not confident about the prospects of getting support from the local Cassa di Risparmio Bank. But to his astonishment, he walked out with credit for a hundred million lire. The now emboldened Budinich felt on firmer ground with the influential Prince Raimondo of Torre e Tasso (in German Thurn und Taxis[4]). His castle further up the coast at Duino had welcomed Franz Liszt, Mark Twain and Rainer Maria Rilke[5], and in the immediate post-war period had been the headquarters of the Allied administration (with the Prince relegated to a tent in the grounds). Donating the land that would become the future home of the Institute, the Prince said 'Trieste is my daughter and this is her dowry'[6]. In 1961, Italy could promise the IAEA that $500 000 for central site and construction costs, together with other contributions, would be available if Trieste were selected as the home for the new centre. Despite the emergence of other strong contenders, it was to be difficult to refuse. Had it not been for the efforts of Budinich, Salam's brainchild would have materialized somewhere else – there had been bids from Denmark, Austria, Pakistan and Turkey, as well as competition from elsewhere in Italy – and would have looked very different.

On their respective fronts, Budinich and Salam were now highly involved in local, national and international politics. The power politics of science had only really emerged in the twentieth century, when scientific development attained an industrial scale, harnessed to war effort. In the Second World War, the highly influential figure of Vannevar Bush foresaw the role US nuclear science would have to play and made sure that appropriate wheels were set in motion. In the ensuing scheme, J. Robert Oppenheimer was chosen to head the scientific development work for the atomic bomb. A theoretical physicist, Oppenheimer had studied in Europe, a necessary move for US scientists prior to 1939, and returned home to found the great US tradition of theoretical physics that went on to become so influential. In the Manhattan project, Oppenheimer had the stature on one hand to command the respect of his talented scientific team, and on the other to safeguard its intellectual integrity, despite enormous political pressures that could easily have blown the huge project off its scientific course. Salam had taken note.

While the Second World War had given US science new momentum, Europe, with the possible exception of Great Britain, was faced with the problem of rebuilding its science from the ruins of the war. Edoardo Amaldi in Italy and Pierre Auger in France became the spiritual fathers of the CERN laboratory in Geneva, the first major European scientific project to arise from the post-1945 rubble. With CERN in place as a role model, Amaldi and Auger went on to propose another new venture, this time to co-ordinate European collaboration in space research, which led eventually to the European Space Agency.

However, the International Centre for Theoretical Physics (ICTP) that emerged at Trieste was not a European venture. Salam's plan was for an international centre for scientists from developing countries, and it made its home in Europe because of the dynamism and imagination of Budinich. A major element in Budinich's plan was that the centre should be headed by a scientific heavyweight with an international reputation. Salam, a scientist of world renown and already an international figure, was the natural choice. Just as the reputation of Ulugh Beg had pulled itinerant Islamic scholars to Samarkand, or the authority of Niels Bohr ensured that scientists continually made a pilgrimage to Copenhagen, so the new centre needed its figurehead. Budinich urged Salam to move to Trieste, where he would be received 'as a Roman Emperor'. Budinich knew what the Centre would bring to Italy, and to Trieste in particular, but Salam alone understood its true purpose and could chart its scientific destiny. Nobody else had such a fervent commitment to science in developing countries, and in his own mind he saw that he would have to be the figurehead of the new centre. In 1964, he became the new institute's director-designate.

In major career moves, Salam had always liked to keep his options open. Thus, when he had left Government College, Lahore, for a lectureship at Cambridge, he was careful to ensure that the door was left open for three years in case he wanted to return. This option was eventually superseded by his nomination as President Ayub Khan's science advisor, which brought new responsibilities. The question for Salam now was how to take on the added responsibility of heading the new centre. At Imperial College, he was not overburdened with teaching, either at undergraduate or graduate level. Paul Matthews, a gifted teacher and administrator, absorbed much of the responsibility there, and his promotion to professor in 1962 had given Salam increased scope. In March 1964, after some exchanges between Salam and Imperial College management, underlined by helpful letters from IAEA Director General Sigvard Eklund, the college initially granted him one year's leave of absence to oversee the creation of the new centre.

In the space of a few years, Salam had made his Imperial College group a key player on the world subnuclear research scene. The college did not want to lose even a fraction of such a valuable piece of research potential so quickly. The eventual solution was that Salam should spend four months of each year at Imperial, and the rest at Trieste. Patrick Blackett, who could always recognize a good thing when he saw one, wrote to Imperial College Rector Sir Patrick Linstead to underline Salam's secondment: 'The [Trieste] Centre is Salam's creation and it might not flourish without him'[7]. Salam enjoyed a good relationship with Linstead. Salam relates how Linstead 'kept a large globe in his office with large pins stuck in it to mark the locations of members of his professional faculty around the world – he took so much pride in what his men could achieve for the world at large'[8]. For Salam, Linstead's pin had originally been stuck in Pakistan. Now he had to insert an extra one in Trieste.

Salam's initial objective at ICTP had been to spend half a day on research, the other half on necessary administration. And this was in addition to his dual responsibilities as a professor at Imperial College and as Science Advisor to Pakistan President Ayub Khan. ICTP's administration was deliberately minimalist. Salam remarked 'The day that the director of a research centre like this stops being a scientist, he's useless. Any fool can administer. People forget that they were made heads of centres because they were doing good science. So they lose their competence. They become manipulators of men just to keep themselves in power.'[9] However, Salam underestimated the effort that would be needed to fund the infant centre, which was to be a continual struggle until the announcement of his Nobel Prize in 1979.

After the provisional approval by the IAEA in June 1963, the agreement between Italy and the IAEA was that the host country would pay some 80% of the running costs of the new centre, providing also buildings, infrastructure and personnel. But Budinich knew that Trieste itself had some catching-up to do. On his arrival in the city after attending the 1962 decision in Vienna, he pointed to the transport and communications improvements that had been implicit in the site criteria for the original IAEA proposal. To accomplish its new role, Trieste's airstrip at nearby Ronchi had to be improved, and the road and rail links with Venice and the rest of Italy had to be upgraded. With the vigorous backing of Ambassador Ortona from the Ministry of Foreign Affairs, this work was soon accomplished. Trieste was no longer held to the rest of Italy by a thread.

The IAEA, in its role as the UN's nuclear police, did not have unlimited cash to support pure research activities, and funding from UNESCO was limited by its initial role of creating new ventures, but not necessarily sustaining them. The initially estimated annual budget for the infant organization was \$525 000, of which the IAEA would contribute a fraction. When the United States had suggested drastically reducing the centre's budget, Isidor Rabi retorted with 'if they wanted a centre for under-developed countries, they would end up with an under-developed centre'. (Several years later, Rabi visited a now booming Trieste centre and apologized to Salam and Budinich for his earlier remarks.) At its 1962 annual conference, the IAEA board of governors voted \$55 000 for the scheme, sardonically described as a 'princely sum' by Salam[10]. There was also \$22 000 from UNESCO. But the offer of \$278 000 from Italy and a promise of purpose-built premises got the idea of an International Centre for Theoretical Physics in Trieste off the ground. The ICTP was voted into existence in 1963, with the formal agreement between the IAEA and the Italian government signed in Rome on 11 October.

The leafy slopes of the Miramare site, seven kilometres north of the city, had now been earmarked for the new centre, but meanwhile temporary accommodation had to be found. The centre's first home was a five-storey twentieth-century building in Trieste's central Piazza Oberdan (named after Guglielmo Oberdan, who in 1882 was arrested before being able to throw a bomb at the visiting Austrian Emperor at an exhibition to mark the 500th anniversary of Habsburg Trieste: Oberdan was tried for treason and hanged). Some deft administrative moves elbowed out some of the building's occupants – a school and branches of local administration – to make space for the new institute.

Salam had now accepted the position of founding Director, with Budinich as his deputy. Salam's one-year leave of absence from Imperial College became a major commitment. The time-sharing with Imperial College was acknowledged in Salam's contract with the IAEA, under which he was paid $2290 per month, free of tax, and that also covered his commuting between Trieste and London. Salam hoped that administration could be run on a shoestring, but he knew that even with the most gifted management, any organization, whether a scientific centre or a commercial enterprise, needs specialist advice and policy guidance. Trieste needed a non-executive board to shape policy and underline his own authority. Its members would have to reflect ICTP's interests and international status. Salam was much in awe of J. Robert Oppenheimer, for his scientific work, for his achievements in science administration, and for his level of culture: Oppenheimer enjoyed poetry, plays and classical literature, and had also studied Sanskrit. He must have relished having a gifted scientist from the Indian subcontinent under his wing at Princeton's Institute for Advanced Study (another had been the Indian mathematical physicist Harish Chandra, who had come to Princeton with his research supervisor, Paul Dirac, in 1947 and subsequently stayed for a second year).

Knowing how influential Oppenheimer could be, Salam had been anxious to involve him as soon as possible in his plans for an international centre. However, the US scientist's stature and influence in his own country suffered greatly in the vicious loyalty purges of the 1950s. Although by now suffering from throat cancer (from which he died in

1967), Oppenheimer accepted Salam's invitation to be a founding member of the Trieste Centre's Scientific Council, where he helped draft the Centre's charter in 1964. The Council, chaired by Manuel Sandoval-Vallarta of Mexico and the Massachusetts Institute of Technology, also included in its early years Aage Bohr, son of Niels Bohr, and who would go on to win the Nobel Physics Prize in 1975; Victor Weisskopf, then Director-General of the CERN laboratory in Geneva; Anatole Abragam from France; and two Soviet scientists, A. Matveyev and V. Soloviev. For his part, Salam represented what had become known as the Third World, the aggregate of nations who had become spectators to the ideological and political confrontation between communism and capitalism.

With even the temporary Piazza Oberdan premises not yet ready, the centre's first scientific venture was a seminar on plasma physics, the science of hot gases, organized in the Jolly Hotel on the Trieste waterfront (which was to become Salam's Trieste home for the next few years). A notable participant was Soviet scientist Roald Sagdeev, one of the youngest ever to be elected a full academician of the USSR Academy of Sciences, and who at that time worked on major Soviet thermonuclear fusion projects. (Later he became Director of the Soviet Space Research Institute, and was an influential advisor to Mikhail Gorbachev.) The topic for the initial Trieste meeting had been deliberately planned. In 1964, the major outstanding problem of applied physics was (and still is) to develop the technology to tame thermonuclear fusion, enabling large-scale power to be obtained from sea water, with no attendant greenhouse gases or radioactive waste. Salam hoped that something would emerge from the meeting to set controlled fusion on a new path. It was also a pilot project to foster collaboration between scientists from east and west at a time when labyrinthine regulations for entry and exit visas made such collaboration difficult.

To help run the new centre, some staff were seconded from IAEA, others were former employees of the Allied Military Government, and the remainder recruited locally. The actual start of operations at the centre was less auspicious. There was little explanation for the sudden appearance of a bunch of introvert physicists in the Piazza Oberdan premises. The Prefetto, the local representative of the central government in Rome, suddenly found himself without a press office, a dilemma that he solved by requisitioning the space housing Trieste's film archive and throwing away the archives[11]. With space limited,

Salam moved in with a few post-doctoral fellows and post-graduate students from Imperial College – 'the Salam boys' – as the nucleus of a research group, compact enough for them all to lunch together at the nearby *Mensa dei Ferrovieri*. Among them were Robert Delbourgo and John Strathdee, who were to become long-term Salam collaborators. At weekends, with more time available, Salam enjoyed eating grilled fish at one of the seaside restaurants, where he would amaze his colleagues by crunching the fishbones, leaving only the head and tail on his plate. Salam kept his researchers continually on their toes, firing instructions at them and almost immediately asking for the outcome. There was a sense of relief among them when Salam departed for his monthly visit to London. However, any relief was short-lived, as he left them much work to do, and he would soon be back.

The Piazza Oberdan offices were bare and dirty in October 1965. The previous occupants had not been in a hurry to move out and refurbishment was chaotic. The first day was an anticlimax for Salam. On the first day in his new fifth-floor office, he wrote 'One cannot even smile. It is cold, raining, miserable. The morning was occupied in trying to get shelves and mats, trying to get the place clean and free of cigarette ends. We discovered there are only three power points per floor, and there will be no heating until 15 November at the earliest. In the afternoon, I, at least, could not stay. I marvel at the uncomplainingness of the others. God forgive me for accepting this place.'[12] However, visitors were soon intrigued and captivated by the institute's pioneer atmosphere, where scientists from Europe, Ghana, Sudan, Ireland, Lebanon and Israel all got along just fine. 'This kind of understanding, solidarity and friendship between people from different cultures is what the world needs most,' said Gerhard Mack, describing his stay at the infant ICTP[13].

By his amazing ability to compress scientific research into whatever time was available in transit or between meetings, Salam could juggle the unpredictable logistics of his new institute with the challenge of new science, work that led in January 1965 to 'The covariant theory of strong interactions', which Salam, ebullient after having overcome the obstacles of getting installed in Piazza Oberdan, heralded as the theory to end all theories of elementary particles. If he had been right, Trieste would have triumphantly become one of the world's leading scientific centres. It would be an inspiring culmination to all Salam's dreams. But it was not to be, and soon he was back at the drawing board.

Despite this setback, Trieste was making its mark on world science in another way. Initially, the institute attracted visitors through seminars and courses. The plasma physics meeting at the Jolly Hotel was the prototype for many more, held at a rate of one every few weeks and covering a wide range of pure and applied physics topics. These meetings, and the visitors they attracted, were one of the spearheads of ICTP's approach. An early and frequent visitor to ICTP was Alfred Kastler, awarded the 1966 Nobel Physics Prize for his work on the optical pumping of atomic energy levels and who spread the message of laser physics via his Trieste lectures. Salam's avatar Paul Dirac, born in 1902, became another regular visitor to the Centre, and would amaze other scientists by swimming across Grignano bay on his way to his office.

But, welcome as they were, these prestigious visitors were not a recipe to counteract the isolation of research workers in developing countries, the frigid isolation that Salam had had first encountered on his return to Lahore in 1951, and the steady haemorrhage of talent that deprived these countries of the skills and know-how needed to improve their lot. For this, the medicine was Trieste's 'associate membership' scheme, under which associates working in developing countries were given travel money and a daily allowance, enabling them to come to the Centre several times over a period of a few years, for periods of up to three months per visit, provided they otherwise continue to work in their own countries. This was Salam's primary 'anti-brain-drain' weapon, directly combating the motivation-sapping isolation that he had himself experienced.

In this way, the number of scientists from developing countries visiting Trieste increased from 60 in the initial Piazza Oberdan years to several thousand. In parallel, a second scheme reflected Budinich's initial plan for a federation of Central European universities under which researchers could commute to Trieste. These federated institutes come from Austria, Czechoslovakia, Hungary, Poland, Romania and Yugoslavia, soon extended to cover South America, Africa and Asia as well as additional centres in Europe. In all these exchanges of visitors, the contact between East and West gave a foretaste of what would only become a political reality in Europe thirty years later. A Soviet visitor wrote of his Trieste experience 'For us, young students, the West was an inaccessible and dangerous, strange world, and the leaders of theoretical physics [here follows a list of Nobel laureates] were more of a legend than real.....I really learned the meaning of complete freedom of thought'.[14]

May 1968 was a milestone month in Salam's life. At the same time as he was making his electroweak presentation to the Nobel Symposium in Gothenburg that would eventually stake his claim to the Nobel Prize, the Institute was moving from Piazza Oberdan in the city centre to its new purpose-built home in Miramare. While the grand new building offered more space, a cafeteria, a library and air conditioning, together with the magnificent surroundings of the castle gardens just across the road, many pioneers regretted having to lose the informality of the early years. The offices at Piazza Oberdan had always been bustling, but a major challenge now was to fill up all the offices in the fine new building. This needed money to fund an increased throughput of visitors.

On 9 June 1968, the new building was formally inaugurated, and the Italian *carabinieri* hoisted the UN flag in front of the building. A major seminar marked the move. For three weeks, 300 of the world's leading experts in particle physics, condensed matter theory, astrophysics, relativity, plasma physics, cosmology, nuclear physics, quantum electronics and biophysics lived together in an attempt to counter the continuing trend towards specialization and generate a new symbiosis in science[15]. Salam gave a rambling talk on the fundamental theory of matter. The real highlight of the programme was an embedded evening session of autobiographical sketches by legendary figures in the twilight of their careers – Hans Bethe, Paul Dirac, Werner Heisenberg, Oskar Klein, Lev Landau and Eugene Wigner – scientists for whom Salam had immense respect. Klein was the only one without a Nobel Prize. (Landau, severely injured in an accident in 1962, had died before the Trieste meeting and his lifestory was described at Trieste by his colleague Eugen Lifshitz.)

Trieste had already shown its value as a corridor where scientists from West and East could meet and mingle. The inauguration seminar called for a substantial contingent of Soviet scientists, and Isaak Khalatnikov, Director of the Landau Institute in Moscow, was asked by the USSR Academy of Sciences to select a team. In those days, it could be difficult for Soviet scientists to attend international meetings. To ensure balanced participation, the organizers of such meetings frequently stipulated attendance quotas for scientists from particular countries or regions. While for most countries there were frequently more candidates than places available, the Soviet quota was often not completely filled. For USSR scientists, a major obstacle was getting authorization to leave the country. Often, scheduled Soviet speakers or delegates would not turn up because of last-minute administrative problems. Even those who got there could have difficulty getting enough hard currency to pay their way, and any peripheral travel required stringent political supervision. Nevertheless, for the 1968 Trieste meeting, the Soviets fielded a strong team. As well as the science, the Soviet delegation also appreciated the ICTP hospitality, which included a car that they drove to Florence. In his talk, A. A. Abrikosov related how he had once encountered a bear while walking alone in the mountains, prompting the usually taciturn Paul Dirac to fire a series of questions[16].

To have so many famous scientists at his centre was a culmination of a dream for Salam, especially as at that time his scientific fortunes were at a low ebb. His U(6,6) theory of strong interactions had been discarded, and his quest to unify weak and electromagnetic forces (of which more later) appeared to have gone totally unnoticed. It was an anticlimax after the carnival atmosphere of the move to Miramare. The scientific despondency was underlined by personal setback. In April 1969, Salam, in New York on United Nations business, got a phone call from London that his father, 78, was critically ill. Muhammad Hussain had returned to Multan in 1959 after a successful cataract operation in London, but was seriously diabetic. Procuring the correct diet in Pakistan had not been easy. Then came heart problems. Salam just managed to get to Karachi from New York in time to see his father brought to hospital from Multan. Shortly afterwards, his father died. For months afterwards, Salam was a broken man, constantly moping, and would not eat properly. Worried, his family sought help from Zafrullah Khan nearby. While accepting that grief is natural and normal, Zafrullah sternly admonished Salam for such exaggerated behaviour. Such *shirk* – idol worship – was un-Islamic. The warning brought Salam to his senses. There was much work to be done.

The ICTP's new buildings were just across the road from Maximilian's fairy-tale castle where Salam had been introduced to Trieste at the 1960 seminar organized by Budinich. When he was at the centre, Salam lived in a small villa near the new building. Initially, his life there away from his family was spartan. Spending most of his time in his office, the house remained in darkness. His diet was frugal, but the advent of freezers enabled him to import supplies of home cooking from London.

Spending so much of his time in Italy, it made sense to learn Italian, even though virtually all communication onsite at Trieste was in English. In his youth, Salam had displayed a talent for languages, acquiring some knowledge, if not fluency, in Urdu, English, Arabic and Persian as well as his Punjabi mother-tongue. Learning Italian did not seem an insurmountable obstacle, but time and motivation were. He attended courses, but frequently dropped out because of other commitments. A Trieste visitor tried a method that he claimed had always worked well – learning the language while walking – 'one step, one word'. Salam occasionally read official speeches written for him, and maintained he was making progress. His Italian colleagues tested this claim by insulting him among themselves in Italian while he was within earshot, without any detectable result. The radio at his Trieste home was tuned to the BBC World Service.

Undeterred by an apparent lack of success of his 1960s research work, Salam's objective was still to share his time more or less equally between administration and science. He had moved on and was now trying to reconcile quantum field theory with cosmological approaches to gravity. Quantum effects are mainly confined to the domain of the infinitely small, atoms and their inner workings, while gravity is the stage of the very large, the motion of stars and their planets. Bringing together these two complementary aspects of Nature was traditionally an extremely difficult task, but one that great minds periodically revisited, the intellectual equivalent of trying to bend a rigid bar so that its ends touched. Perhaps a reconciliation of these opposite extremes could help find a way through the morass of renormalization problems. Even if the chasm between gravity and the quantum world looked unbridgeable, at least the people on each side could try to shout across at each other. Specialists from the two camps came together at Trieste in 1970, and one outcome was a paper by Salam, Christopher Isham and John Strathdee that showed how infinities could be suppressed in gravity-modified quantum electrodynamics[17]. Salam, as a proponent of the field theory school, had once bet with cosmologist Hermann Bondi, a research colleague of Fred Hoyle, that important features of general relativity could be reproduced using field theory techniques. It was demonstrated in a PhD thesis by Michael Duff, now the Abdus Salam Professor of Theoretical Physics at Imperial College. There is no record of whether Bondi paid up[18]. Salam reported on these developments at a physics meeting in Amsterdam in 1971 in a session chaired by Martin Veltman. It was in this carefully stage-managed setting that Gerard 't Hooft announced to the world that field theories with massive particles can be renormalized. Salam had been awarded the first talk – 'I let Salam talk about his baloney', said Veltman[19]. After the Amsterdam meeting, field theory was to abruptly change direction, now with the unification of weak and electromagnetic interactions as its main objective.

For the other half of Salam's time, funding was still a major preoccupation. During the first years of ICTP's operations, most of the money came from the Italian government. Pleas to IAEA and to UNESCO did not go unheeded, but the IAEA's annual budget in the early 1960s did not permit generous support to any of the organization's peripheral activities. Salam and Sigvard Eklund had to expend a lot of energy to secure additional Trieste funding from other sources, notably the Ford Foundation in the US. UNESCO had been one of the founding ICTP sponsors, but only in 1970 did UNESCO support for the centre's actual operations increase to become comparable with that of the IAEA. Ironically, this highly welcome funding increase would almost destroy what ICTP stood for. From the outset, ICTP's objectives had been to overcome political barriers. Strategically placed at the centre of Europe, it attracted scientists from the Western and Eastern blocs, providing a useful intellectual crossroads that supplemented its main role as an incubator for scientists from developing countries. However after 1970, the centre's closer ties to its UNESCO foster-parent were to bring unexpected political problems.

The United Nations, set up in the aftermath of the Second World War, reflected the balance of power at that time, with the victorious Allied Powers having a permanent presence with veto rights in the main decision-making arena, the Security Council. To channel its efforts in specific areas, the UN oversees specialist organizations, such as the IAEA, the World Health Organization, the International Monetary Fund and UNESCO. The open forums of their General Assemblies are frequently a sounding board for contemporary international opinion. In an atmosphere clouded by the outcome of the 1973 Yom Kippur war between Israel and its neighbours, a series of anti-Zionist resolutions by the UNESCO General Assembly effectively excluded Israel from the organization's operations. Whatever its motivation, this sudden 'politicization' of UNESCO's primary cultural role was criticized by prominent intellectuals in the US and Europe. The US, the traditional ally of Israel, roared with indignation. Piqued by other UNESCO developments, the US soon withdrew from the organization (as did the UK and Singapore). With ICTP now perceived as UNESCO-sponsored, and with its close contacts to Third World countries that had supported the new resolutions, for the first time icy political winds began to blow through the Trieste corridors. Then, Salam's institute was officially boycotted by the US and by Israel, and their scientists turned their backs on Trieste-related activities[20], seriously disrupting the smooth running of workshops and courses. Especially harrowing for Salam was a 1975 event planned to mark the sixtieth birthday of British cosmologist Fred Hoyle, who almost 30 years previously had been his tutor at St. John's College, Cambridge, and who had steered him from an undergraduate career in mathematics to one that covered physics as well. When so many invited scientists announced that they would not attend any meeting at ICTP, Hoyle's birthday event had to be rescheduled for neutral Venice. To try to control the damage, Salam offered to make his former pupil, Yuval Ne'eman, a 'corresponding member' of the Trieste Centre. A consummate politician, Ne'eman refused. UNESCO soon lifted its sanctions on Israel, and operations at Trieste began to run normally again, but it would be many years before the USA would rejoin the UNESCO fold.

By the mid-1970s, Salam had realized that long-term funding was going to be a major problem. He approached oil-rich Iran, at that time still ruled by the Shah, but without success. In April 1975, the Directors General of UNESCO and the IAEA invited Salam to Paris to discuss future funding for the ICTP. Salam asked that staunch supporters and impressive figures, in the person of CERN Research Director General Leon Van Hove and Nobel Physics laureate Alfred Kastler, should also attend. The current IAEA and UNESCO funding was insufficient for Trieste's needs, and in addition the Institute was feeling the full icy blast of the UNESCO-related Israeli/US boycott. Salam was on very thin ice. Rather than whingeing about the boycott, he instead displayed unswerving loyalty when eloquently pleading with his traditional sponsors. UNESCO and IAEA heard him and pledged to extend their support for the centre for another term. With this achieved, Salam then activated his contacts all over the world to pressure their IAEA and UNESCO delegations. In mathematics, it is essential to pay scrupulous attention to detail. A single tiny error can make a whole calculation crumble. Salam orchestrated his push for funding like a mathematical equation. In 1977, both the UNESCO and IAEA contributions were substantially boosted.

With Italy no longer the largest single contributor to ICTP funds, the next objective was to pressure that country, and Salam's Nobel Prize in 1979 provided an ideal opportunity. Salam says that he actually threatened to leave and take the centre with him unless the Italian government substantially increased its contribution[21]. In 1979, Italy again became the largest single contributor to the ICTP coffers. In 1981 Italian Foreign Minister Giulio Andreotti visited the centre, accompanied by prominent Italian physicist Antonino Zichichi, who announced to a happy throng crowding the lecture hall that the Italian contribution to ICTP was being increased to $5 million, at which point Andreotti stood up and announced 'I am sorry to correct my friend Antonino, but the contribution henceforth will be $10 million.'[22] The sums mentioned in fact bracketed several years, but thanks to Andreotti, whose Foreign Ministry could turn on financial taps in both the foreign and development areas, the Italian contribution increased exponentially during the 1980s, exceeding $1 million in 1981 and $7 million in 1986.

With its funding assured, ICTP became a role model for new ventures, both by Salam and in a wider context. In 1963, writing in the Bulletin of Atomic Scientists[23], Salam recalled how an eleventh-century physician in Bokhara had broadly classified his medicines into those that remedy diseases of the rich, and those that cured the poor. The same is still true, said Salam: developed countries worry that their inhabitants are too obese and have heart attacks, other nations face famine, starvation and malaria. One part of humanity has to live under the shadow of nuclear weapons, while another suffers vitamin deficiencies or cannot afford effective pharmaceuticals. The root of all this modern dichotomy, said Salam, was an excess of science on one side and a deficiency on the other. All his life he strove to redress this imbalance, which in more recent years has been accentuated by environmental problems and energy resources. The success of his scheme at Trieste pointed a way forward.

The purpose of the Trieste centre was to give scientists from the Third World research opportunities that they would not otherwise have. Apart from individual scientists, what about the countries themselves? In October 1981, during a meeting of the Pontifical Academy of Sciences in Rome, delegates from Third World countries rued the absence of a voice for science in their respective countries, either because the country was too small, or because the scientific activity was easily overlooked. One of the delegates at Rome was Mambillikalathil Menon. After being a key member of Cecil Powell's group at Bristol that had discovered new particles in cosmic-ray experiments in the late 1940s and early 1950s, 'Goku' Menon returned to India to work at the Tata Institute for Fundamental Research in Bombay, eventually becoming Director General of India's Council for Scientific and Industrial Research. From this platform, he perceived how the voice of science in developing countries was not heard. While the US or the UK listened when its influential National Academy of Science or its Royal Society spoke, Menon pointed out to Salam that there was no such spokesman for Third World science. The delegates in Rome signed a joint statement and went away, leaving Salam with a baby in his hands. Thus was established the Third World Academy of Sciences (TWAS), a new guardian of science and scientific development for developing countries, 'the South'[24].

Initially nurtured from within ICTP itself, TWAS could stand on its own feet after a life-giving transfusion of $1.5 million from the Italian government. With Salam still a great believer in the role the United Nations had to be seen to play in international science, the Academy was formally opened by UN Secretary-General Perez de Cuellar at Duino Castle on 5 July 1985. In his address, Perez de Cuellar pointed to three regrettable trends: the continuous refinement of the technology of destruction; the degradation of the environment; and the prevalence of poverty, ignorance and disease in developing countries. He called on scientists in developing countries to redress the balance. The existence of a new focus for Third World science should help, he said.

The pomp and spectacle of the inauguration of the new institute underlined for Salam the need to find someone to run it. He didn't want another job: he had enough on his plate already. Mohamad H. A. Hassan, a Sudanese physicist with a doctorate from Oxford, had first come to see the new Institute at Trieste in 1974 after purchasing equipment for his father's soap factory in Italy. Salam, seeing another promising scientist starved of research opportunity, encouraged Hassan to become an associate, and make short annual visits to Trieste. From this, Hassan became Salam's right-hand man during the creation of TWAS, the new organization's executive secretary. The Annual Meetings of TWAS soon became the springboard for new initiatives, addressing critical fields of science and technology. Salam saw existing schemes that could be used as models. The Consultative Group on International Agricultural Research (CGIAR) had been spawned in the mid-twentieth century through collaboration between Mexican agriculture and the Rockefeller Foundation, and taken forward by the realization that developing countries would starve if nothing were done. Other specific agricultural problems were being attacked by the International Rice Research Institute in the Philippines, and the International Maize and Wheat Improvement Centre in Mexico. The results could be spectacular, with record harvests and improved conditions for farmers and workers alike.

Salam envisaged similar schemes for other areas of applied science, funded by the World Bank, which had supported CGIAR. Visiting the World Bank in 1991, the now infirm Salam was no longer the firebrand who had galvanized the IAEA thirty years earlier. He was advised to go away and seek support from individual governments first. Others would have cut their losses and written off the original idea. But the guardian angel of the Italian government stepped in again, this time to underwrite a new research and training centre for high technology and new materials, chemistry, and the study of Earth Sciences and the environment.

While ICTP remained the prototype for other new ventures, by Salam and by others, its commitment to the minutiae of theoretical physics prevented it from being a global management hub. To provide a logical focus for new ventures, Salam conceived of the idea of an International Centre for Science and High Technology as an umbrella organization. After the creation of the Third World Academy of Sciences, these ideas received new impetus. Salam's ideas were now feeding each other. Via the now traditional route of UN and Italian government support, new centres were set up under the banner of the United Nations Industrial and Development Organization (UNIDO). The International Centre for Genetic Engineering and Biotechnology (ICGEB) was sited in Trieste and New Delhi, while the International Centre for Pure and Applied Chemistry (ICC), the International Centre for High Technology and New Materials (ICTM), and the International Centre for Earth and Environmental Sciences (ICE) were all centred in Trieste. Salam played a key role in the establishment and administration of these bodies.

Other Salam Third World interests were less specific: Food and Disarmament International, The Fund for Physics in Developing Countries, the International Commission on Peace and Food, the

International Foundation for the Progress and Freedom of Science, and Scientists for Human Development. The establishment of all these centres and networks of centres increased Salam's visibility as a champion of science in developing nations. In 1989 he was invited by former Tanzania President Julius Nyerere to represent science in the South Commission, established by the United Nation's G-77's Third World coalition under Nyerere's chairmanship. As the spokesman for science in Nyerere's South Commission, Salam ensured the implications for science and technology in the commission's influential 1990 report. The following year, he was invited to the World Bank's Consultative Group on International Agricultural Research, where he pushed for the establishment of a network of international centres of excellence along the ICTP model in various fields of applied science and technology. This proposal was initially greeted by economists with the scepticism that had met his proposals to the IAEA in the early 1960s.

ICTP's core business was theoretical physics, especially of the particle, nuclear and plasma varieties. But in response to continuing demand from developing countries and from UNESCO, the scope of this science gradually extended to include condensed-matter physics and mathematics, with ten-week introductory courses. Later this was extended still further to cover physics and the environment, high technology (including lasers, fibre optics, microprocessors and materials science), industry, medical and biophysics. While most of these courses were held in Trieste, some were organized in Africa, Asia and South America, there to dovetail with local efforts.

Salam's original idea had now expanded to a whole web of institutions, sometimes resembling more a tangle, but he knew which were the important threads and could control it. However, the widening scope of all this science in turn led to the city of Trieste itself becoming a pole of attraction, with new institutes such as the International School for Advanced Studies (Scuola Internazionale di Studi Superiori Avanzati – SISSA), with Budinich as founding director after his formal retirement from ICTP. SISSA, the first Italian institute of higher education to award the PhD degree, was a reflection of ICTP that sidestepped any potential political problems of being a Third World research centre. Through the influence and energy of Budinich, a major synergy naturally developed between the ICTP family of institutions and the University of Trieste. The science at Trieste expanded far beyond the particle theory that had first attracted visitors to Miramare Castle.

A by-product of the huge atom-smashing machines built all over the world for subnuclear physics research was the short-wavelength radiation given off as particle beams were whipped round and round. This 'synchrotron radiation', which sapped the energy of the stored beams, was initially looked on as a waste, but soon scientists realized that if this peripheral radiation could itself be concentrated into beams, it could be used to 'X-ray' atomic and molecular structures in all kinds of materials research. A powerful synchrotron radiation machine equipped with suitable instrumentation soon became a necessary feature for all major industrial countries. Italy's 260-m circumference machine, ELETTRA, began operation in 1993 in Trieste's new AREA science park. Science in Trieste had come a long way since Salam's theory group first moved into converted offices in the Piazza Oberdan.

The establishment of ICTP and its support from IAEA and UNESCO sparked interest at the heart of the United Nations organization. In 1969, in his introduction to the Annual Report to the General Assembly, UN Secretary-General U Thant proposed the idea of an international university, 'because his attention had been drawn to the work being done by individuals to establish institutions of learning with an international character'[25]. Salam became a member of the UN study groups and foundation committee for this new venture. Salam privately hoped that his centre could be the nucleus of the new university[26], but with generous funding from Japan, the new UN University was duly established on 6 December 1973 as an international community engaged in research, advanced training, and the dissemination of knowledge related to pressing problems of human survival, development, and welfare. From its Tokyo headquarters, the University oversees a worldwide network of research and postgraduate training centres.

At the centre of his self-spun web, Salam used a variety of management styles. He was a shrewd judge of people, and knew those whose help he could request, those he could flatter, coax or cajole into helping, those he could issue orders to, and those he could bully. If he had judged wrongly, and his target tried to slip away by giving an apologetic excuse, he would wave his hand around and proclaim 'Do you think it was easy to build this Centre? Where do you think the money to run it comes from? Do you suppose it runs itself?'[27]

Despite all this scientific enterprise, and the fulfilment of the ambition he had first glimpsed in Lahore, there was one crushing disappointment. Salam was proud of his religion and of the great heritage of

Islam. In his Nobel Prize address, even before sketching the science for which he earned the award, he outlined the history of Islamic science and its importance for the pre-renaissance world, for ardent seekers of knowledge like Michael the Scot. But then Islam and science had gone their separate ways and somehow lost each other. Salam's ambition was to reconcile them. 'There is no question but today, of all civilizations on this planet, science is weakest in the lands of Islam,' he declared. 'The dangers of this weakness cannot be overemphasized since honourable survival of a country depends directly on its strength in science and technology in the condition of the present age'[28]. In 1973, to coincide with an Islamic summit in Lahore, he proposed that the Pakistani government take up his proposal for an Islamic Science Foundation to promote science in Islamic countries. Its plan was to build up a new infrastructure for modern science, with specialized institutes and authoritative scientists, and a scheme, along the lines of the Nobel Foundation, to recognize contributions.

The proposal led to a loose association of Muslim scientists '*Ummat-ul-Ilm*' (Community of Science) that first met at Trieste at Salam's suggestion in 1980. (He had written its constitution in 1975[29].) Speaking at a UN symposium in Kuwait in 1981, Salam used his most grandiose words to remind fellow-Muslims of the glory that had been Islamic science, when intellects like Al Haytham, Ibn Sina and Al Biruni had widened the view of the world centuries before Newton[30]. Support in his home country was vital, but with Salam's voice hardly discernible in Bhutto's Pakistan, his proposal fell on deaf ears and became sidelined. With the exception of the Gulf States, awash in oil revenue, support did not emerge. In 1990, his powers waning, Salam said 'after 25 years preaching, some funds have become available from the Gulf. If we can obtain similar funds for Muslims in general, this may make a big difference to the prospects for physics in Islamic countries.'[31] It was not to be. Salam travelled tirelessly to further his cause. Subsequently, with his Pakistani passport now branding him as an Ahmadi heretic, he could no longer travel to Saudi Arabia, the most important potential supporter for any pan-Islamic venture. At Salam's insistence, Pakistan President Zia-ul-Haq agreed to raise this issue at major international meetings. Despite the fine words[32], and incessant lobbying and travelling, the proposal was largely ignored. 'It may have been more charitable not to have deceived ourselves by its creation,' Salam wrote later[33]. Even the little money that was raised became a target for venom from orthodox mullahs in Pakistan: 'Salam takes six million dollars from oil-rich countries . . . and gulps it down. Immensely pleased he is that a large sum has been received, unshared, to convert Muslim youth into Qadianis'[34].

After the successes of the Ayub Khan era, Salam's efforts in his native Pakistan foundered. When his Nathiagali College for 'Physics and Contemporary Needs' was established in 1976, the hope was that it would develop into a national centre along the lines of Trieste. However, Salam's lobbying of the National Bank of Pakistan did not produce the same results as Budinich had achieved for Trieste at the Cassa di Risparmio. Subsequent efforts targeted Karachi industrialists, where the hope was that Pakistan could produce a scientific centre like the famous Tata Institute in India. After Salam had been silenced by illness, a Commission on Science and Technology for Sustainable Development in the South (COMSATS) was finally established in 1994 with its headquarters in Pakistan.

Despite the publication of Salam's book 'Renaissance of Sciences in Islamic Countries'[35], there was no renaissance. Salam's imagination and enthusiasm continually exploded like a firework, but there were no Muslim equivalents of Sigvard Eklund and Paolo Budinich to run with these schemes and ensure the necessary local support for them to take root and flourish. Any hope for generous patronage of Islamic science remained a forlorn one. In his later years, Salam spent much time and energy touring corridors of power, soliciting support for these schemes. His reputation ensured a polite reception and sometimes some seed money, but nothing got off the ground. The disappointment was underlined by the stark contrast to what had been achieved at Trieste, To Salam's anguish, Islam and modern science remained largely irreconcilable.

REFERENCES

1. Monfalcone is 20 km further north
2. Budinich, P., *L'arcipelago delle meraviglie*, (Rome, Di Renzo, 2002)
3. Now part of Croatia, off Rijeka
4. The dynasty had had a monopoly on stagecoaches, hence the word 'taxi'
5. Duino was where the Austrian physicist Ludwig Boltzmann committed suicide in 1906.
6. Vauthier, J. *Abdus Salam, un physicien* (Paris, Beauchesne, 1990)

7. Imperial College London archives, KP/13
8. Lai, C. H., Hassan, Z., (ed.), *Ideals and realities, selected essays of Abdus Salam*, (Singapore, World Scientific, 1984)
9. Lai, C.H. Hassan, Z., (ed.) 1984 reprinted from an interview published in the UNESCO Courier.
10. Salam, A., *Physics and the excellences of the life it brings*, in *Pions to quarks*, Brown, L., Dresden, M., Hoddeson, L., (ed.) (Cambridge, CUP, 1989
11. de Greiff Acevedo, A., 'The International Centre for Theoretical Physics 1960–1979: Idealogy and Practice in a United Nations Institution for Scientific Cooperation and Third World Development', Dissertation submitted for the Degree of Doctor of Philosophy, University of London, Imperial College of Science, Technology and Medicine, December 2001
12. Salam, U.,*Trieste and my father*, in Hamende, A. M. (ed.), *Tribute to Abdus Salam*, (Trieste, ICTP, 1999)
13. Mack, G. *Conformal symmetry*, in Ali, A., Ellis, J., Randbar-Daemi, S. (ed.) *Salamfestschrift* (Singapore, World Scientific, 1994)
14. Filippov A. T., *Soviet theoretical physics and the ICTP*, quoted in Hamende, A. M., *A Guide to the early history of the Abdus Salam International Centre for Theoretical Physics*, (Trieste, Consortium for Development and Research, 2002)
15. Salam, A., Fonda, L., (ed.), *Contemporary physics: Trieste symposium 1968* (2 vols), (Vienna, IAEA, 1969)
16. Khalatnikov I. M., *Our great contemporary*, in*100 reasons to be a scientist*, (Trieste, ICTP, 2004)
17. Isham, C. J., Salam, A., Strathdee, J., *Infinity suppression in gravity-modified quantum electrodynamics*, Physical Review D3 1805–17 (1971)
18. Duff M.J., *Twenty years of the Weyl anomaly* in Ali, A., Ellis, J., Randbar-Daemi, S. (ed.) *Salamfestschrift* (Singapore, World Scientific, 1994)
19. Crease, R., Mann, C., *The second creation, makers of the revolution in twentieth-century physics* (New York, Macmillan, 1986)
20. de Greiff, A., *The politics of non-cooperation: The boycott of the International Centre for Theoretical Physics*, Osiris, **21** p.86–109, 2006
21. Salam, A., *A life of physics*, in Cerderia, H. A., Lundqvist, S. O., (ed.) *Frontiers of physics, high technology and mathematics* (Singapore, World Scientific, 1990)
22. Hamende, A. M., *A great scientist and administrator*, in Hamende, A. M. (ed.), *Tribute to Abdus Salam*, (Trieste, ICTP, 1999)
23. Salam, A., *Diseases of the rich and diseases of the poor*, Bulletin of Atomic Scientists, **Vol XIX, No4**, April 1963
24. Schaffer, *D., TWAS at 20, a history of the Third World Academy of Sciences*, (Singapore, World Scientific, 2005)
25. www.unu.edu
26. Budinich, P., *Fulfilled and not yet realized dreams* in Hamende, A. M. (ed.), *Tribute to Abdus Salam*, (Trieste, ICTP, 1999)

27. Qadir, A., *Servant of peace* in Hamende, A. M. (ed.), *Tribute to Abdus Salam*, (Trieste, ICTP, 1999)
28. Salam, A., foreword to Hoodbhoy, P., *Islam and science* (London, Zed Books, 1991)
29. Lai, C. H., Hassan, Z., (ed.), (1984)
30. Salam, A., (ed Dalafi, H. R., Hassan, M. H. A.) *Renaissance of Sciences in Islamic Countries*, (Singapore, World Scientific, 1994)
31. Salam, A., (ed. Dalafi, H. R., Hassan, M. H. A, 1994)
32. These papers and speeches are collected in Salam, A., (ed. Dalafi, H. R., Hassan, M. H. A, 1994)
33. Lai, C. H., Hassan, Z., (ed.), (1984)
34. http://www.alhafeez.org/rashid/ludhianvi/abdussalam.html
35. Salam, A., (ed. Dalafi, H. R., Hassan, M. H. A, 1994)

12

Electroweak

An anthem for the 1960s was Bob Dylan's poetic 'The Times they are A-Changing'. During those restless years, the oppressive shabbiness of the immediate post-war period evaporated, and the world glimpsed a new future. In France, newly affluent and stripped of its colonial past, students occupied university campuses and rebelled against inherited anachronisms. Following their example, European workers staged general strikes. In the US, young people rallied around the standard of civil rights, roared out their disapproval of their nation's ruthless war in Viet-Nam, or explored ways of dropping out. In pragmatic Britain, young people spurned politics and blew new wealth on music and gimmicky fashion. Boys grew their hair long and girls wore their skirts short. London became Europe's crucible of innovation. With a new generation of researchers swept by the tides of these developments, science too was in transition.

After setting up his group at Imperial College, Salam had attained his intellectual cruising speed. He increasingly appeared wherever there was intellectual action, like a moth around a flame. While this activity was admired by some, his continual flitting to and fro across the research stage amused others, and was criticized by those who felt that research should not be so capricious. In the push and shove of the research race, Salam clung obstinately to the leading group. This sometimes dissipated his energies and blurred his objectives. But whatever the research fashion, he had one deep scientific conviction to which he invariably returned after all his other dalliances

In 1907, Albert Einstein had what he called 'Der glücklichste Gedanke meines Lebens' (the happiest thought of my life). The force of gravity that keeps our feet firmly on the ground has only a relative existence. A person falling freely does not feel their own weight: only when the fall is broken does gravity reassert itself. After having this 'happiest' thought, Einstein was silent for three and a half years as he searched for a mathematical framework in which to mount it. He found it in the work of Georg Bernhard Riemann, who had developed a new calculus of space, building up lines, sheets and volumes of any shape and size by adding together infinitesimally small parts. Instead of constructing a framework of girders, it was like building a sandcastle from individual grains of sand. Using Riemann's geometry, Einstein's 'general relativity' showed how gravitation could be understood as the geometry of space.

Looking at how general relativity had been formulated, the German mathematician Hermann Weyl objected to what he saw as sloppy thinking. Einstein had assumed that it was possible to compare the lengths of the same ruler at two different places. Instead, said Weyl, such a ruler, or 'gauge', can only be used at one time and place. The measured length at some other time and place is given by the equations of the theory. If the theory could be constructed in such a way that it was transparent to the way the ruler was used, then it was 'gauge invariant'. Using this principle, Weyl believed that he had not only made the theory more rigorous, but had also found a way of welding gravity and electromagnetism together in a single unified picture of Nature. According to Weyl, electromagnetism was also geometry: the same forces that held the stars in their heavenly courses would also control the microscopic structure of the atom. It was a compelling idea. If he were right, Weyl had upstaged Einstein. Motivated more by the search for truth than any jealousy, Einstein soon showed Weyl to be wrong, but, impressed by such a grand objective, for the rest of his life stubbornly tried to succeed where Weyl had failed. Einstein was left intellectually stranded, a museum piece at Princeton's Institute for Advanced Study.

Physics sets out to explain Nature through a set of underlying assumptions, or axioms. Progress comes by finding which axioms have the most general implications. One is the apparently innocuous requirement that the same experiment should always give the same result, no matter where or when in the Universe it is conducted. This seemingly trivial premise in fact implies that energy and momentum have to be conserved. Such immutabilities – when an entire event-structure shifts in space and time without affecting the outcome – are called 'global' invariances or symmetries. But there are more subtle kinds of such symmetry. For a hosepipe laid flat on the ground, any cross-section looks the same, no matter where it is cut from. The pipe – a long cylinder – is symmetrical around its long axis. But a hose is not rigid. If one end is twisted slightly, the hosepipe still looks the same from the outside and each cross-section still looks symmetric, but only at the cost of introducing stresses in the pipe's material. These forces result from distortions in the hose, and its cylindrical symmetry has to be carefully specified. Such 'local' symmetries introduce strains and torsions.

Weyl's ambition to extend the principle of gauge invariance eventually did strike home, but in a different context. The new setting was quantum theory, whose equations describe things that cannot be directly measured, and are only indirectly related to the physical world that we see. In their own invisible space, the quantum equations carry mathematical baggage that can disappear when quantum results sublimate into visible reality. However, one fingerprint of the quantum formulation is electric charge, which, even when the quantum content has been distilled off, remains visibly conserved because of an underlying gauge invariance in the mathematical formulation of the theory.

In what was to be a lifetime commitment, Salam espoused quantum gauge theory at the beginning of his research career. His first forays with Paul Matthews took the electromagnetic field as a model and tried to build a counterpart theory of the nuclear field. For a variety of reasons these initial attempts failed. They also took no account of newly discovered subnuclear labels[1] that behaved like electric charge. Once physicists knew that subnuclear particles carried such additional labels, perhaps an extended gauge theory of subnuclear processes could be constructed with the new charges playing a role analogous to that of the electric one. The first to try in the early 1950s was Wolfgang Pauli, now more than 50 years old, but still a driving force. With his theory closely modelled on electromagnetism, he soon came across a problem. His gambit threw up a new set of particles that were supposed to carry the subnuclear forces. Analogous to the electromagnetic photon, they had no mass. There was no shortage of exotic subnuclear particles, but none of them had zero mass. In fact most of them were quite heavy.

Also fascinated by the idea of such a gauge theory was C.N. Yang, who was soon to be, with his Chinese colleague T. D. Lee, the pioneer of left–right symmetry violation in weak interactions, but who in 1954 was working at Brookhaven with Robert Mills. Younger and less conventional than Pauli, they were less embarrassed about having a theory with particles without mass somehow generating a strong nuclear force. In 1954 Yang presented these ideas at a seminar at the Institute for Advanced Study, Princeton. In the audience lurked Pauli, who knew about unwanted massless particles and immediately pounced. Each time Yang mentioned the particles, Pauli reiterated his objection. Acknowledging the mass problem, Yang admitted that he did not know how to solve it and wanted to put it to one side for the moment.

'That is not sufficient excuse,' thundered Pauli.

Yang, completely confused, sat down and the seminar room fell silent. Oppenheimer, the Institute's Director, stood up and said 'I think we should let Frank proceed'[2].

Pauli never published these ideas, but despite Pauli's initial objections Yang and Mills did, and their 1954 papers became a pointer for future developments of quantum gauge theories. Another who had visited these ideas was Ronald Shaw, Salam's student at Cambridge, whose 1955 PhD thesis included a chapter 'Invariance under general isotopic spin transformations', work that Shaw had done the previous year. In it, Shaw acknowledges that his work was not published and that the idea was subsequently proposed by Yang and Mills[3]. To the annoyance of those already annoyed by his research antics, Salam frequently referred to what everyone else called 'Yang–Mills theory' as 'Yang–Mills–Shaw theory', but the more diffident Shaw preferred anonymity. After Salam got to know Pauli better after the 1956 neutrino mass saga, Pauli and Shaw compared research notes. In a 1957 letter from to Salam (by then at Imperial College), Pauli added a postscript 'I am looking forward to hear from Mr Shaw about the Yang–Mills problem'[4]. The rash of unwanted massless particles deterred most people the way they had deterred Pauli and field theory quickly went out of fashion. The Yang–Mills paper was initially overlooked, but eventually went on to become the focus of field theory research, the starting point for several important new developments. Shaw's parallel ideas remained buried in his Cambridge thesis.

As field theory went out of fashion, theoretical physicists left adrift quickly climbed aboard a mathematical lifeboat – 'dispersion relations'. Anxious to remain at the research forefront, Salam briefly did too, working with Walter Gilbert at Cambridge and later with Paul Matthews at Imperial College, but he never forgot his commitment to gauge theory. In 1959, at the height of field theory's obscurity, Salam first met the influential Soviet theorist Lev Landau at that year's major physics meeting, held in Kiev (just three years later, Landau was awarded the Nobel Physics Prize for his work on the superfluidity of liquid helium). In 1956, Salam and Landau had independently proposed that a massless neutrino could account for left-right asymmetry in beta decay. Salam relates how Landau, wearing a flamboyantly coloured shirt, introduced himself at Kiev. He asked Salam 'Aren't you a believer in field theory? Aren't you ashamed of yourself?' The puzzled Salam enquired why. 'I have just shown that the Hamiltonian for electrodynamics in field theory is zero'[5]. Landau's physics-speak meant that the theory should collapse. He had discovered the existence of 'ghost' particles, shards of negative probability that overturn the logic of cause and effect. These ghosts were to plague field theory, already discredited by massless particles, for many more years. But resolute field theorists like Salam were secretly undeterred. They had seen how renormalization had cleansed their theories of unwanted infinities. Surely something would come along to take care of massless particles and Landau's ghosts.

The Yang–Mills approach provided a useful template for those resolute field theorists. They knew that a conserved quantity reflects some deeper symmetry, in the case of electric charge the quantum version of Weyl's gauge invariance. Other charge-like quantities (such as isotopic spin and strangeness) are conserved in the strong interactions that produce exotic sparks of subnuclear particles, but not in the weak interactions that then extinguish the sparks. As exotic subnuclear particles decay, their additional charges are not conserved, but the outcome somehow appears to retain a memory of them.

Salam had taken his first hesitant steps towards classifying subnuclear particles at Cambridge with his student John Polkinghorne. To extend these ideas, he acquired new research partners, at Imperial and further afield. A new centre of attraction opened up in 1955 when enough buildings were available at CERN, the new European Centre in Geneva, to house a few young theorists. Among the first to be hired were Bernard d'Espagnat and Jacques Prentki, from the French Centre National de la Recherche Scientifique (CNRS) in Paris, who worked on the adventurous kind of physics that Salam liked. In 1956, when Salam invented his massless neutrino explanation for the left–right dependence on weak interactions, one of his first ports of call was the temporary building near Geneva airport where d'Espagnat and Prentki worked. In his 1979 Nobel lecture, he vividly described how they worked alongside a gas ring 'on which was cooked the staple diet of CERN – *entrecôte à la crème*'. A nice story, but difficult to accept. Few Frenchmen would be prepared to cook their own lunch and eat it at their desks when there were several acceptable bistros nearby. In any case, CERN was already far too big to be fed from a kitchen with a single gas ring. More interesting for Salam was to see at first hand how the preparations for the new European laboratory were shaping up. Certainly CERN was going to be a major player on the European scientific scene, if not the world, and a prototype for international scientific collaboration. Beside the theory of subnuclear particles, by 1956 Salam had the germ of other ideas in his mind.

In 1957, an imaginative effort by Salam and his CERN friends d'Espagnat and Prentki set out to classify particles and their various kinds of interaction. Their picture used two spaces, one four-dimensional, the one that Salam had looked at earlier with Polkinghorne, the other with three dimensions. The idea was that strong nuclear interactions were described by the invisible four-dimensional space, where another charge (isotopic spin) is conserved. This contains several three-dimensional subspaces, which apply to electromagnetic and weak nuclear forces, but where isotopic spin is not conserved. Physics at that time was very confused: in many cases, the quantum charges assigned to the various particles were wrong or could only be guessed at. In the days before Gell-Mann and Ne'eman had shown how to assign particles to SU(3) families, these attempts by Salam and his colleagues look old-fashioned and cumbersome, analogue solutions in a pre-digital age.

Salam continued these investigations with John Ward in 1960 after they had run into each other at Robert Sachs' summer institute at the University of Madison, Wisconsin. In view of the brevity of the paper (just one page) and its very specific content, it was probably one example of Ward's allegation that Salam added his name to a paper without asking permission. Salam enjoyed the contacts he made at Madison. One was the Brazilian physicist Jayme Tiomno, At the Madison summer institute in 1961, he and Tiomno analysed the apparent quantum charges in the latest results on exotic particle decays. Tiomno had studied with John Wheeler, Feynman's first teacher at Princeton. With Wheeler, Tiomno had discovered in 1949 how the different types of weak interaction all operate with the same strength. Returning to Rio de Janeiro, Tiomno was keen to develop the new Brazilian centre for particle physics, and had invited Feynman to visit Brazil. As well as exposing Brazilian students and scientists to the newest developments in science, the visit also showed Feynman the attractions of Copacabana Beach and increased his resolve to leave his permanent post at Cornell in upstate New York for somewhere where the sun shone more often. Soon afterwards Feynman took a job at the California Institute of Technology (Caltech) in Pasadena. After Salam took up his job at Imperial College, Tiomno, always keen to attract vigorous young researchers to Brazil, invited him to South America. Salam describes how his path there crossed that of Feynman. 'I was giving a lecture on . . quantum field theory. I was surprised to see Feynman at the end of the room. As soon as I began my talk, Feynman started to interrupt with basic questions like "what is a particle?". I replied "Feynman, these are fundamental questions which we do not understand . . . . Questions which nobody can answer yet. Will you kindly answer them yourself or shut up", at which point Feynman did.'[6] The contacts Salam made in South America were stored away in his social databank to be used later when setting up his international centre in Trieste.

Sometimes new ideas emerge more or less simultaneously in several places, ideas whose time has come. At the beginning of the 1960s, it was the turn of spontaneous symmetry breaking. This had been discovered in several other branches of physics, and the hope was that adapting these ideas to subnuclear particles might help get field theory out of its impasse of massless particles. By now, physicists had discovered symmetries like SU(3), which appeared to carry an important message. The particles that fitted into the SU(3) families bore a family resemblance, but were nevertheless easily distinguishable by their different masses – the symmetry was not exact.

In spontaneous symmetry breaking, a system that is otherwise completely symmetric falls into an outcome that is not symmetric, usually because of some inherent instability. An example is a pencil finely balanced on its point – a very unstable situation, and the slightest touch makes the pencil fall in some arbitrary direction, breaking the original symmetry. Salam gave a nice example of spontaneous symmetry breaking: imagine a round table set for a banquet, with a table napkin N between each place setting, X. The round table appears as a symmetrical loop XNXNXN . . . . ., and as each diner sits down, he or she in principle can pick up a napkin with either the left or the right hand. However, once the first diner chooses a napkin, the rotational symmetry shifts from the table to the diners, and all the other diners have to follow suit. The exact symmetry of the table setting has been spontaneously broken by diners with a right hand and a left hand. Elsewhere in physics, such ideas had been used to help understand magnetism and superconductivity.

Jeffrey Goldstone had studied physics at Cambridge and had been interested initially in the behaviour of nuclear matter. After leaving Cambridge, he had research posts at Copenhagen, CERN and Harvard, where his research focus shifted to particle physics and he began looking at field theories with spontaneously broken symmetries. Finding an example of such a theory that naturally gave massless particles, in a 1961 paper 'field theories with "superconductor" solutions' he conjectured that all such theories would naturally inherit massless particles. This was not helpful. The massless particles that Yang and Mills had introduced were bad enough. Now, spontaneous symmetry breaking, greeted as the saviour of the zero-mass dilemma, appeared instead to provide more. At the summer physics institute at Madison, Wisconsin, in 1961, where Salam was chatting to Tiomno about the weak decays of exotic particles, Goldstone had long talks about these developments with Salam and with a young researcher called Steven Weinberg.

After his first degree at Cornell, Weinberg had gone to Niels Bohr's Institute for Theoretical Physics in Copenhagen for his research debut, returning for his PhD work to the US, where he looked at the implications of renormalization for weak interactions. During this time, he worked out a tough theorem that extended the early renormalization results of Dyson and Salam. Although active in many different aspects of particle theory, Weinberg, like Salam, retained a deep commitment to field theory. Salam first met Weinberg, then a young postdoctoral researcher, at the University of California at Berkeley in 1960. After learning of Goldstone's conjecture of massless particles, Weinberg, with a prestigious Sloan Fellowship, wanted to follow up the idea, and in October 1961 arrived to spend a year in Salam's group at Imperial College. Here, Ne'eman was learning about group theory and Salam was developing his own ideas on particle families. In London, Weinberg's contact with Salam was intermittent, but they also communicated with Goldstone at Harvard, and the three finally published an elaborate proof that any such theory with global symmetry would have to contain massless particles, now called 'Goldstone bosons'[7]. Goldstone, the distant partner in the collaboration, later said 'Basically I didn't write a single word of that paper. They wrote it. But in fact some of the results in fact were things we had discussed at length in Wisconsin'[8]. In his Nobel lecture in 1979, Weinberg recalled that he was so discouraged by these massless particles that he included a quote from Shakespeare's 'King Lear' in the paper he wrote describing his work with Salam and

Goldstone – "Nothing will become of nothing: speak again". The *Physical Review* 'suggested' removing the imaginative quote before publication. The 1962 paper was a milestone *en route* to a distant destination, one that physicists could not yet discern through a fog of confusion. Goldstone, Salam and Weinberg overlooked that their theorem only applied to global symmetries, where the symmetry is the same everywhere. They had not looked at the more subtle implications of local symmetries.[9] In addition, physicists were also confused between symmetries, like SU(3), that were approximate, and genuinely spontaneously broken ones.

With all the dramatic developments around SU(3) and the discovery of so many exotic particles, in the early 1960s it was the strongly interacting particles that were the main focus of attention. After the success of the omega-minus prediction, strong interactions and quarks clamoured for attention. The weak nuclear interaction was less glamorous than its strong counterpart. After the parity violation episode of 1956, it had not made any more newspaper headlines. Using beams of selected particles, physicists used the strong interaction as a forge to fashion exotic subnuclear species and fill SU(3) families. Each discovery of a new particle seemed to guarantee a physics professorship. The weak interaction promptly destroyed the new particles, but it was the weak interaction's signature that identified them. What exactly was this destructive force? There were clues. Atoms, held together by electromagnetism, could easily fray at the edges and lose electrons. The weak force, operating only inside nuclei, deep inside atoms, released beta particles – electrons. Why should the same products – electrons – emerge from two very different worlds? Could there be some link between electromagnetism and the weak force?

A clue had been planted long before. Oskar Klein, born in 1894, the youngest son of Sweden' s first rabbi, Gottlieb Klein, was one of the most imaginative physicists of the early twentieth century. Physics is littered with his name – Kaluza–Klein theory, the Klein–Gordon equation, the Klein paradox, the Klein–Nishina formula. Klein's mind continually blazed like a firework, firing off brilliant ideas in all directions. These ideas were uncontrollable, and often out of step with the way science was progressing. Had he stuck with one, or even a few, of these ideas, the name of Oskar Klein would surely have become even more famous. At a meeting in Kasimierz, Poland, in 1938, Klein suggested that beta decay could be mediated by particles that played an analogous role in weak interactions to that of the photon in electromagnetism. He called them 'electrophotons'. But unlike massless photons, which can roam anywhere, they were heavy, and thus could not get far under quantum conditions. With the clouds of war gathering once more over Europe, and with Klein's suggestion being published in a Polish journal at about the time that Nazi troops were entering the country, the idea went unnoticed for almost twenty years.

In the late 1940s, new forms of weak interaction had been discovered. Tiomno and Wheeler and others pointed out how these new kinds of reactions seemed to behave in the same way as ordinary beta decay, and resurrected Klein's idea of electrophotons. One of the proponents, T. D. Lee, suggested calling the heavy carriers W, for 'weak', but this idea fell flat, and lay dormant for the best part of a decade, until the revelation early in 1957 that left–right symmetry was violated revived interest in weak interactions. One who recalled the idea was Julian Schwinger, one of the architects of quantum electrodynamics. Schwinger's ambitious 1957 paper 'A theory of the Fundamental Interactions' was submitted to *Annals of Physics* on 31 July, after *Physical Review* had objected to some of Schwinger's terminology. Journals were becoming sensitive to the literary efforts of their contributors. Schwinger tried to pull together everything that was then known about strong, weak and electromagnetic interactions. This knowledge was incomplete, and his effort went wide of the mark, but to bring together the weak and electromagnetic interactions he proposed augmenting the electrically neutral photon of electromagnetism with a pair of oppositely charged particles whose ponderous heaviness ensured that their weak interactions had a much shorter range than the photon. Schwinger talked about the idea at physics meetings, a voice in the crowd.

Becoming a physics research student in those days was difficult. Subnuclear physics in the post-war period was developing so fast that textbooks were out of date before they were published, as Salam had discovered. Graduate students had to plunge headfirst into a turbulent morass, struggling to learn the confusion of their teachers. As they tried to make sense of their allotted tasks, new revelations, like parity violation in weak interactions, continually blew away whatever wisps of comprehension had developed. In all this turmoil, there was no shortage of problems to attack, but in assigning research problems to their students, professors had to be sure that the students would have something to show for their efforts. Salam had warned Ne'eman of these problems.

When parity violation was uncovered, Sheldon Glashow, who had gone to the same New York City high school as Steven Weinberg, was a graduate student at Harvard, listening intently to everything that Schwinger said. Glashow wanted to become Schwinger's research student, and after having convinced the master that he was worth taking on, was assigned a problem. Glashow's contemporaries were given very specific problems, exact calculations using tried and tested procedures. Schwinger was still convinced that the weak interactions and electromagnetism were deeply interrelated. His 1957 paper had given some suggestions. But the objective he gave the student Glashow was unusually open-ended – to seek such an interrelation, and explore the consequences. It could have been a disastrous introduction to research in the hands of a less capable student. Glashow did not solve the problem, but explored enough of the consequences of Schwinger's suggestion to earn a PhD degree. Afterwards, Glashow wanted to spend a year doing research in the Soviet Union. This was an administrative as well as a scientific challenge, and Glashow moved to Niels Bohr's Institute in Copenhagen to await a Soviet visa, which never came.

Another who read Schwinger's 1957 paper was Abdus Salam. His work with d'Espagnat and Prentki at CERN had suggested the very idea of a three-dimensional charge space that Schwinger had needed. Salam now took the idea further. Because extra particles were needed from the start, he claimed that the carriers of the weak force arise 'naturally', and that gauge invariance in this charge space generates the required forces, in the style that Yang and Mills had proposed for strong interactions. Salam's massless neutrino broke the symmetry and ensured parity violation for weak interactions. In the paper[10], Salam acknowledged the pioneer work of Schwinger, and that he had heard about Glashow's work. He stressed that the two charged carrier particles of the weak force need to be heavy, and make an uneasy triplet with the massless photon of electromagnetism. 'We propose to come back to this problem in a subsequent paper', 'we' being Salam and John Ward, who also appeared to sign the paper. Later, Ward described this episode – 'Abdus went ahead anyway and published a premature effort with me (*nolens volens*) .... I would certainly have objected to this.'[11]

It is less clear if Ward would have objected to his name appearing on a paper based on ideas that he had discussed with Salam at the Madison summer institute in 1960 and that appeared in a paper 'On a Gauge Theory of Fundamental Interactions', which, still in pre-SU(3) days, attempted to wrap all three interactions – strong, weak and electromagnetic – in a single parcel. It acknowledges parallel work by Glashow. Parity violation is still missing and has to be put in by hand. It is a messy paper, stilted with the pre-occupations of the time, and concludes 'in all this work, parity violation for weak interactions remains a complete mystery'[12]. In his 1979 Nobel talk, Glashow described some of this work as a 'remarkable portent' of what was to come, but few could understand what Salam and Ward were trying to do. At the time the paper had been written, their feelings were not reciprocated. Salam and Ward knew that trying to renormalize their new theory would be difficult, but proceeded anyway. Meanwhile, Glashow claimed to have produced a renormalizable version. Salam and Ward, then the world's acknowledged authorities on renormalization, were amazed that a then unknown upstart youth claimed to have done what they could not. Piqued, Salam subsequently showed that Glashow's claim was wrong, with the result that for some time he never read anything more by Glashow, which he later admitted was a mistake[13].

By now Glashow was installed in Europe and once more productive. In September 1960 he submitted for publication an updated version of his doctorate work, still following Schwinger's deep conviction that the weak and electromagnetic forces had to be related in some way. This time, Glashow added an extra ingredient. To make everything hang together, ensuring that the weak interactions broke left–right symmetry but electromagnetism didn't, he added an additional neutral particle. The weak force was now mediated by a charged triplet of heavy particles – positive, negative, and neutral – the new neutral one being a new form of weak interaction, shadowing the electromagnetic photon in some way. Until then, all weak interactions had involved a shuffling of electric charge. Glashow was demanding a form of weak interaction that had never been seen[14]. He sidestepped this objection by assuming that the extra neutral particle was much heavier than its charged counterparts. 'The baby was lost with the bathwater,' he said in his 1979 Nobel lecture. Salam was still not reading what Glashow wrote, and had anyway been diverted by the strong-interaction developments centred on SU(3), trying to backtrack after having initially discounted Ne'eman's efforts. For these attempts to construct new gauge theories, a few specialists were valiantly trying to ensure such theories would be renormalizable, capable of supplying reliable quantitative results. Here, the mysterious neutral form of the weak interaction opened new possibilities for magical cancellations that could wipe out troublesome infinities.

In the meantime, Salam's life had turned a corner. In Cambridge, Salam had become isolated from religion. Where there is no mosque, a Muslim is free to offer prayers wherever he is. There is no guilt feeling, but on the other hand the sense of community and spirit instilled by a fervent congregation is not there to nurture and sustain faith. In addition, the sojourner is surrounded by new influences and immersed in a climate that can dampen religious fervency. In the early 1960s, visitors to Salam's office reported having seen a bottle of Scotch. Alcohol is forbidden to Muslims. However, Salam could have meant it as a gesture of hospitality: in those days even non-smokers could keep an elaborate cigarette box to proffer to visitors. In 1962, Salam took his wife and parents to Mecca to perform *Umrah*, the small pilgrimage. Involving a single lap of the *Ka'aba*, this can be done at any time of the year, and involves much less organization and effort than the elaborate full pilgrimage, the *Haj*. The experience nevertheless impressed him deeply. Every Muslim is supposed to make the full *Haj* once: making *Umrah* does not absolve a believer from the responsibility of making the full pilgrimage. But it was to be Salam's only trip to Saudi Arabia. When the Ahmadi movement to which Salam belonged was declared heretical by Pakistan in the 1970s, the passports of Pakistani Ahmadis henceforth carried a stamp declaring that their bearers were heretics, and that automatically barred them from entering Saudi Arabia. It would be a painful blow for someone whose faith was a major pillar of his life. But in 1962 this was still a cloud on the horizon in an otherwise bright sky. Salam returned from Arabia with his faith resurgent, and with fresh motivation.

Plunging back into science, the revitalized Salam joined with Ward again for a 1964 paper with a tighter synthesis of electromagnetic and weak interactions. (The peripatetic genius of Ward had lodged at Johns Hopkins University, Baltimore, but had still not found his definitive career niche. In 1963 he had been offered a job at Oxford, a place where he had never been happy, and that he turned down 'for obvious reasons'[15]. He was soon to depart for New Zealand.) Ward, passing through Imperial that summer, admits this synthesis was 'done right'. After their impenetrable 1961 paper, the second opens on a major chord – 'One of the recurrent dreams in elementary particle physics is that of a possible fundamental synthesis between electromagnetism and weak interactions.' At this point Salam and Ward link back to earlier work – their own 1959 paper, which Ward did not approve of, Glashow's 1959 paper with its topheavy neutral particle, and Schwinger's 1957 suggestion. 'The idea has its origin in the following shared characteristics,' they continued, going on to list the same points that had been noticed by Oskar Klein, now underlined by the fact that a much wider range of weak interactions had become known. Electromagnetism acts as a current with a definite direction. So does the weak interaction – a 'weak current'. Then came the bad news. The strengths of electromagnetism and the weak force are very different, which quantitatively implies that the carrier of the weak force would have to be about 140 times heavier than the proton. For this 'outrageous mass', Salam and Ward could offer no explanation. They just shut their eyes and went ahead. Opening them again, they had the spectre of the mysterious new 'neutral current' of the weak interaction, 'the minimum price one must pay'[16]. While Glashow had pushed this under the carpet by making its carrier very heavy, Salam and Ward did it by rendering it weaker than weak.

Apart from its physical insight, this paper is remarkable because in 1964 almost everyone was busy with SU(3) and quarks, including Glashow, who appeared to have forgotten about his earlier attempts to synthesize weak and electromagnetic effects. Ward was very excited about it. J. D. Jackson of the University of California, Berkeley, was in Europe that year and later recalled a dinner with Ward, who enthusiastically described his latest efforts with Salam, saying that they were in the verge of a major breakthrough, if only they could think of the right mathematical transformation. Scribbling down the formalism on a paper napkin, he asked for assistance, but by this time too much Gevrey-Chambertin 1958 had been consumed for Jackson to be helpful, however, he kept the wine-stained napkin in case[17].

The spontaneous symmetry-breaking ideas that people hoped would make sense of approximate symmetries had been sidelined after Goldstone, Salam and Weinberg had 'proved' that they would be plagued by unwanted massless particles. But not everyone was convinced. Robert Brout in Brussels had noticed that the special case of forces with infinite range had escaped the Goldstone–Salam–Weinberg net[18]. He phoned Salam, who promised he would look into the problem, but never did. Others noticed more special cases. Salam's former student, Walter Gilbert at Harvard, in his last dalliance with particle physics in 1964 before crossing the bridge irrevocably to molecular biology, pointed out that one of these objections was flawed. Gilbert's paper in *Physical Review Letters* was read late at Edinburgh University, because the journal only came via surface mail. One who read it there was Peter Higgs. By using gauge theory and relativity, Higgs could overcome Gilbert's objection, and quickly sent off a contribution to the Editor of *Physics Letters*, Salam's friend Jacques Prentki at CERN in Geneva. A few days later, Higgs had developed a full formulation of how particles could acquire mass, and sent this to *Physics Letters* too. It was turned down. Judging that he had been too obscure, Higgs added some suggestions about how it could be linked to the currently fashionable SU(3) picture and sent it to the US journal *Physical Review Letters*, which did accept it, providing that he added a note to say that an independent formulation had been developed in Brussels by Brout and François Englert. Higgs had been unaware of this.

Gerald Guralnik was a student of Walter Gilbert at Harvard, working on spontaneous symmetry breaking, one of Gilbert's last physics students before he quit physics for molecular biology. Travelling through Europe en route to a year at Imperial College, Guralnik met up with Gilbert in Italy, where he was now giving lectures in biology, for a final discussion on spontaneous symmetry breaking[19]. When Guralnik arrived at Imperial, most of the attention was focused on SU(3) developments, but some people remembered the deep allegiance to field theory that Salam had instilled in the group, in the same way that Schwinger's field-theory fervency inspired Harvard. In fact, Harvard and Imperial were among the rare places in the world where field theory was still held in esteem (another was Utrecht in the Netherlands). At Imperial, Tom Kibble and particularly Ray Streater had distanced themselves from the clamour of the SU(3) bazaar. Streater's speciality was the axiomatic mathematics of field theory, and was trying to prove the Goldstone boson theorem more rigorously, while Kibble, a perfectionist, wanted to make sure that Yang–Mills field theories were well behaved.

Kibble had joined Salam's group at Imperial in 1959 when it was still in the Mathematics Department. Like Salam, Matthews and Delbourgo, he had been born in British India, where, like Paul Matthews, his parents were missionaries. By some amazing quirk of fate that must have amused Salam, Kibble's father was Professor of Mathematics at Madras Christian College, the same institution where Matthews' father taught English. Kibble studied physics at Edinburgh, where his research supervisor was John Polkinghorne, who was in his first university post after being Salam's research student at Cambridge. With Kibble, Guralnik continued to work on his relativistic spontaneous symmetry-breaking ideas, now with another US visitor, Richard Hagen. The result was a joint paper that paralleled the results of Higgs and Brout and Englert. Kibble later extended these results to field theories of the Yang–Mills breed.

The subtle distinction between an approximate symmetry and a spontaneously broken one was not yet widely understood, and as fast as they were proposed, these new ideas of spontaneous symmetry breaking were continually being seized upon by researchers trying to explain why SU(3) symmetry was only approximate, with the particle masses within SU(3) families much more disparate than those of the proton and the neutron. One was Steven Weinberg at MIT, who wanted to construct a model of the strongly interacting particles and their masses. He was giving talks about it. 'At some point in the fall of 1967, I think while driving to my office, it occurred to me that I had been applying the right ideas to the wrong problem,' he said in his 1979 Nobel lecture. Weinberg was stressed out. He had taken leave of absence from his regular job at the University of California, Berkeley, and moved to Boston so that his wife could study at Harvard Law School. They had just moved house, and Weinberg had taken responsibility for ferrying his small daughter to and from nursery school. When he arrived at the office that morning, the penny dropped. He quickly shifted the focus of his idea, and where before he had been going round in circles, 'now everything was easy'[20]. The field particles he had to focus on were not strongly interacting ones: they were the massless photon and the particles alleged to be responsible for the weak force. There were four of them: three would be given a mass by the new Higgs mechanism, the fourth, automatically massless, was the electromagnetic photon. The theory contained three interaction strengths – the well-known weak current that stirred up electric charge; the as yet unknown weak neutral current that didn't; and the electromagnetic force. To relate these to those actually seen needed a mixing parameter, now known as the Weinberg angle. Weinberg published 'A Model of Leptons' in 1967[21], then forgot about it.

Salam learned from Kibble how the ideas he had put forward with Ward in 1964 could now avoid massless Goldstone particles. Salam now spoke of the 'Higgs–Kibble' mechanism, a phrase that jarred with

Kibble, aware of the extensive pedigree of spontaneous symmetry breaking. Instead of SU(3), Salam was now pre-occupied with his U(6,6) dream for a fully relativistic version of SU(3), and had also become immersed by the day-to-day problems of his new institute in Trieste, whose funding always seemed to skate on very thin ice. Nevertheless, in theory workshops at Imperial and lectures to graduate students in 1966–7, Salam outlined the new developments. Research students, now struggling with long calculations in Salam's U(6,6), could not fathom why he was now looking at weak interactions. They were being coached in strong interactions and did not know about Salam's 1964 work with Ward. One who did know was Robert Delbourgo, one of the architects of U(6,6), who was organizing theory talks and seminars at Imperial. He told Salam about Weinberg's paper. Distracted by his complicated matrimonial life (of which more later), and by his Trieste centre emerging from its Piazza Oberdan chrysalis, Salam had taken his eye off the scientific ball that he had held aloft the longest – what would become electroweak theory. He saw that he had to react quickly.

The Nobel Foundation regularly held 'Nobel Symposia', covering topics in which Nobel prizes had been earned. From 19–25 May 1968, some 30 distinguished physicists were invited by Niels Svartholm to a meeting on elementary particle physics in a mansion just outside the Swedish city of Gothenburg. As the delegates arrived, many of them compared notes about the student unrest at their respective campuses. In the middle of the meeting, before making his talk, Salam had to jet back to London to iron out funding problems for his International Centre in Trieste, which was moving from its temporary accommodation in the city centre to a fine purpose-built edifice. Scientists from all over the world had been invited to its formal inauguration on 9 June.

Salam's presentation at Gothenburg began with the same words that he had used with John Ward in their 1964 paper – 'One of the recurrent dreams in elementary particle physics is that of a possible fundamental synthesis between electromagnetism and weak interactions'[22], going on to list the common characteristics of the two forces. Salam used his now habitual particle assignments, as had been used in his 1964 work with Ward. These now appear whimsical, but there was the inevitable mixing parameter to construct the physical weak neutral current and electromagnetic current from the two neutral currents of the theory. After linking to his earlier work with Ward, Salam continued 'The material I shall present today…was given in lectures (unpublished) at Imperial College. Subsequently I discovered that an almost identical development had been made by Weinberg, who apparently was also unaware of Ward's and my work.' The problem was to sidestep any unwanted massless Goldstone particle, which 'sits like a snake in the grass ready to strike'. His stated objective to extend the ideas to all particles was still in abeyance, and his final words were 'The hadron problem is still unsolved,' by which he meant bringing in quarks and SU(3). He would return to this later with a new collaborator, Jogesh Pati. As a professional of renormalization, Salam knew that his model would have to be tractable if it were to be of any use. At the meeting, the discussion after Salam's presentation focused on such technical points. After one question, Salam replied 'If you ask me for a dictionary to the conventional theory, ... I haven't got this dictionary worked out yet.' Salam then apparently forgot about his renormalization 'dictionary'. At Gothenburg, the physics content of Salam's talk went unremarked. Gell-Mann did not mention it in his summary talk.

Weinberg's perception had come in a flash of inspiration while driving to his office: Salam's had been the result of a long and arduous quest, but had almost been overlooked in the frantic buildup to getting his new institute in Trieste installed in its definitive home. After publication of the two papers, there was not even a whisper of 'Eureka'. The Science Citation Index, which records a 'hit' each time a scientific paper is referred to in print, gives the following score for Weinberg's 1967 paper: 1967, 0; 1968, 0; 1969, 0; 1970, 1. Weinberg was not even referring to his own paper, and the single hit in 1970 was Salam! It was difficult to achieve a worse score, but Salam's version, hidden deep in the proceedings of the Nobel Symposium, did just that, and Salam meanwhile had focused his attention elsewhere. Weinberg and Salam's proposal also went unmentioned in summary talks at major international meetings that traditionally highlighted new developments and research trends. If there was infertile ground to fall on, these papers had found it. Salam's physics fortunes had hit rock bottom. His U(6,6) theory of strong interactions had been jettisoned and now his attempt to unify weak and electromagnetic effects had gone completely unnoticed.

Experiments had diligently looked for weak neutral currents, irrespective of what any theory said. An unfortunate arithmetical error in 1963 had led to the limit for the appearance of neutral currents being set far too low[23], so nobody had bothered looking further. For a long time, the accepted wisdom was that neutral currents did not exist, and that anyone who thought otherwise was a fool. Salam and Weinberg's work had also been overlooked because in the late 1960s field theory was still out of fashion. However, such fickle trends had not permeated the University of Utrecht in the Netherlands, where Leon Van Hove had presided over a theoretical physics group oriented towards 'statistical mechanics', describing the collective behaviour of matter composed of unstructured assemblies of constituents, which can be anything from atoms to sand grains or galaxies of stars. University science education in the Netherlands had suffered because of the Second World War, and major figures had emigrated to the US, but the tall figure of Van Hove appeared as a saviour at Utrecht.

There, in 1966, Martin Veltman began teaching theoretical particle physics, then an unusual topic in that country. Isolated from his fellow researchers and from the rest of the world, he did not notice that field theory had fallen from favour, and continued diligently to fashion mathematical models. Veltman's quest was to find new domains for quantum field theory. The general consensus had been that Yang–Mills theories with massive particles would not be renormalizable. Salam had claimed this to be so in 1960[24], and influential people believed it for many more years. However, there was a growing feeling that something had been lost along the way. Quantum theory, based on probabilities, has to ensure that the sum of all these probabilities is 100%. This conservation of probability (the 'unitarity' of the theory) had been overlooked. In 1969 Veltman showed that, under certain conditions, he could restore credibility to at least a corner of field theory. The formalism was not hopelessly divergent. Salam was among the first to notice[25], remembering his earlier work and his vow the previous year at Gothenburg to develop a renormalization dictionary. But Veltman had not been aware of what Salam had done. 'I did not read this lecture until much later..... The proceedings of the Nobel Symposium is not a particularly popular channel of communication.'[26] Veltman sensed that what had been intended to be 'proofs' of non-renormalizability were instead conjectures. He fervently believed that counting infinities and trying to reconcile them was not enough: the whole theory had to be tractable and self-consistent, like the imposing edifice of quantum electrodynamics.

In 1969 Veltman was assigned a young student, Gerard 't Hooft, whose uncle, Nicolaas Godfried van Kampen, was also Professor of Theoretical Physics at Utrecht. 't Hooft's great-uncle, Frits Zernike, had won the 1952 Nobel Physics prize for his invention of the phase-contrast microscope. 't Hooft wanted to do particle theory, and soon he was working with Veltman on models of massless particles in Yang–Mills field theory. The next step was described by Veltman. 'Somewhere in the autumn or winter of 1970–71 we walked together from one building of the institute to another. . . . I said something like "All this stuff about massless theories is very nice, but if we only had one renormalizable theory of massive charged vector bosons, no matter how far removed from reality." ['t Hooft] answered "I can do that". This moment is grafted in my brain, as I almost ran into a tree.' The student 't Hooft had gone much further than his teacher had dreamed. Veltman said 'Write it down, we will see. And he did, and we saw.'[27] They had at last discovered how to construct a Yang–Mills edifice that gave particles with mass, and was renormalizable.

In the summer of 1971, the big international conference in particle physics was at Amsterdam. Veltman carefully stage-managed the session on field theory. Salam stood up and showed how he had tried to avoid infinities by bringing in gravity, then went back to his seat in the front row. T. D. Lee gave another talk. Then Veltman, bubbling with glee, introduced the unknown 't Hooft. After the presentation, the word went around quickly. A novice student working at an obscure university had developed a framework in which weak and electromagnetic interactions could be unified in a way that gives full renormalizability. It revealed 'Weinberg and Salam's frog to be an enchanted prince'[28]. This rediscovery of work that had been done several years before and its renormalization redemption was the big physics sensation of 1971. After years of wandering aimlessly in the desert, field theory was suddenly restored to its former glory.

One of the first to react was Weinberg. He knew that the unification idea demanded the existence of neutral currents, a new form of weak interaction that somehow jogged particles without affecting their electric charge. The arithmetical error that had long pushed them under the carpet had meanwhile been uncovered. It was time to look for them again.

The ultimate laboratory for weak interactions was Pauli's neutrino, which carried no electric charge and made its presence felt only through the weak interaction. After discovering these particles in 1956, physicists had learned how to fashion them into beams, rays of weak interaction that could shine through tons of steel and concrete, but perhaps shake a few atoms en route. Weinberg saw that the effect predicted by his theory was just low enough to have escaped detection. A huge new neutrino experiment was being built as the flagship venture of a new laboratory just outside Chicago, with a 6.4 kilometre ring of magnets to whirl protons to unprecedented energies. The laboratory was called Fermilab, in honour of Enrico Fermi, the Italian Nobel prize immigrant who had rechristened Pauli's particle as the neutrino. The initial aim of Fermilab Experiment 1 was to look for the heavy electrically charged carrier of weak interactions, called the W particle, long sought but never seen. Carlo Rubbia, a young Italian physicist then working at Harvard, had suggested a massive upgrade of the experiment, and the suggestions from Weinberg refocused its objectives. Neutral currents, formerly in disrepute, were now top of its agenda.

Across the Atlantic at the European CERN laboratory in Geneva, another neutrino experiment was also getting ready. With all the additional electronic gadgetry of Fermilab Experiment 1, and the energy of CERN's beams only a tenth that of Fermilab, it looked like a scientific David versus Goliath. Cast in the role of David at CERN was a 1000-tonne subnuclear camera called 'Gargamelle', after the mother of Rabelais' gluttonous giant Gargantua. Looking for a wispy fingerprint, and with the Fermilab wrestling with unfamiliar new apparatus, in 1973 the Europeans got there first. For the first time since Cecil Powell had glimpsed unfamiliar tracks left by cosmic rays in 1947, Europe had made a major discovery in subnuclear physics. In his Nobel lecture, Salam recalled the news: 'I still remember Paul Matthews and I getting off the train at Aix-en-Provence for the 1973 Conference [on high-energy physics] and foolishly deciding to walk with our rather heavy luggage to the student hostel where we were billeted. A car drove from behind, stopped, and the driver leaned out. This was [Paul] Musset [one of the leaders of the Gargamelle team]. He peered out of the window and asked "Are you Salam?" I said "Yes". He said "Get into the car. I have news for you. We have found neutral currents". I will not say whether I was more relieved for being given a lift because of our heavy luggage or for the discovery of neutral currents. At the Aix-en-Provence meeting that great and modest man [André] Lagarrigue [the leader of the Gargamelle team] was present and the atmosphere was that of a carnival.'

With the missing neutral currents now found, experiments became quantitative rather than qualitative, measuring the strength of the newly discovered effect. The theory could say nothing about that. The final confirmation came in 1979, in an elegant experiment at a two-mile-long machine at the Stanford Linear Accelerator Centre, California, using the world's highest energy beam of electrons. Normally electrons bounce off matter by electromagnetic effects. But occasionally they can bounce by neutral current effects. These, being weak interactions, are left–right sensitive. If the electrons in the beam are arranged with their spins pointing in the same direction, like tiny magnets, this left–right asymmetry might be detected, intertwining delicately with the dominant electromagnetic effects. There it was, just one part in 10 000. The first announcement outside Stanford was at a meeting at Salam's centre in Trieste in June 1978. Several months later, Salam addressed an international physics meeting in Tokyo on 'Unification and New Ideas'. He spoke of the 'electroweak' force – the synthesis of electromagnetism and the weak nuclear force, the fulfilment of what Oskar Klein had proposed in Kasimierz forty years before. It was the first time the word 'electroweak' had been heard in public.

Soon afterwards, the Royal Swedish Academy of Sciences set in motion the process that would lead to the announcement of the Nobel Prize winners for 1979. Following the usual procedure, some 3000 confidential forms were sent out. In principle, nominations can be submitted by Swedish and foreign members of the Royal Swedish Academy of Sciences (Salam had become a member in 1970); by Members of the Nobel Committees for Physics; by previous Nobel Laureates in Physics; by professors of physics at the universities and institutes of technology of Sweden, Denmark, Finland, Iceland and Norway, and the Karolinska Institutet, Stockholm; by holders of corresponding chairs in universities or university colleges selected by the Academy of Sciences, all with a view to ensuring the appropriate distribution over the different countries and their seats of learning; and by other scientists from whom the Academy sees fit to invite proposals[29]. Early in 1979, the received forms were screened and a preliminary list of candidates selected. After these were assessed by specially appointed experts, the Nobel Committee submitted its report and recommendations to the Academy, who chose the winners in early October by a majority vote.

Salam had already accumulated prizes and distinctions, but he yearned for the Nobel. While he had been a student at Cambridge, during the dying echoes of the Rutherford era, British scientists had won the Nobel Prize for Physics four times in five years (Appleton in 1947,

Blackett in 1948, Powell in 1950 and Cockcroft and Walton in 1951), and he had witnessed the impact this had made, both at Cambridge and in the country as a whole. Afterwards, the Physics Nobel became dominated by US scientists as that country inherited the driving role.

The Atoms for Peace conferences in which he had been a scientific secretary in 1955 and 1958 became the platform for a prestigious 'Atoms for Peace' award, financed by a donation from the Ford Motor Company, in recognition of important developments in issues related to peaceful uses of atomic energy. The first award went to Niels Bohr in 1957. In 1968, after Salam had established his centre for theoretical physics in Trieste, the award was to be shared by Salam, IAEA Director General Sigvard Eklund and Princeton physicist Henry DeWolf Smyth, US delegate to the IAEA[30]. This award brought great pleasure to Salam's father Muhammad Hussain, just one year before his death. In his acceptance speech, Salam groomed his presentation of the romantic thirteenth-century figure of Michael the Scot, soon become a regular feature of Salam's stage act. After the close encounter with Nobel Prize research after his neutrino work in 1956, Salam had told his father that his ultimate goal was now a Nobel science prize. The now-ailing Muhammad Hussain replied 'I know you are more keen on prizes for science, and do not care so much for [this one] for peace, but tell me what power was it that told me to give you the name Abdus Salam (Servant of Peace).'

Salam had been aware of the frustrating proximity of the Nobel ever since his formulation of the massless neutrino explanation for left–right asymmetry in weak interactions had overlapped with the award of the Nobel to Lee and Yang in 1957. After this, he carefully studied whatever he could find on the sociology of Nobel awards, and circulated updated summaries of his publications to those likely to be canvassed by the Royal Swedish Academy of Sciences. As the years went by, he saw the award go to theorists he knew – Lev Landau, Richard Feynman, Hans Bethe, Murray Gell-Mann, Burton Richter, Sam Ting, Philip Anderson, Nevill Mott, ... He could see that patience was needed: Hans Bethe had to wait for thirty years. On several occasions, Paul Dirac, as a Nobel laureate, had put Salam's name forward to Stockholm. So had Bethe. Momentum built up after the key discovery of the neutral current in 1973. Salam's colleague Paul Matthews wrote to Stockholm to explain how Salam had lectured on the unification idea at Imperial College in 1967, prior to its publication in the obscure proceedings of the Nobel

Symposium. Some people who heard 'electroweak' uttered at Tokyo were involved in the Nobel decision over the following months.

Salam was in London when he heard the news, almost simultaneously from Stockholm and from IAEA Director General Sigvard Eklund, who was closer to the Nobel nerve centre. Immediately, messages of congratulation began to pour in from all over the world. His immediate reaction was to go to the London Mosque at Southfields to say a traditional prayer of gratitude for blessings bestowed. In these prayers, he remembered his father and prayed for him. Congratulation messages continued to arrive. Others felt Salam's role had been exaggerated. John Ward, who knew the value of his collaboration with Salam, felt slightly bitter – 'what you gain on the swings you lose on the roundabouts'[31]. Commenting on Salam's publication policy, Ward added 'The conflict between premature publication and fear of being scooped was now endemic. The more expert players developed a technique of the two-way bet. Obscure journals could be used to prove priority if need be and conveniently forgotten otherwise.'

A few days after the Nobel news, Salam spoke at an Executive Board Meeting at UNESCO's headquarters in Paris. He began his speech by acknowledging the laboratories whose experiments had confirmed the electroweak theory, beginning with CERN, where UNESCO had played an important role in the foundation of the organization in the early 1950s. Salam then turned to the parable of Michael the Scot, the figurehead thirteenth-century itinerant scholar who would figure prominently in Salam's Nobel speech in Stockholm two months later. After sketching over his own career and that of his Trieste institute, where UNESCO had become a major partner, he came to money. Salam had always taken pride at Trieste in dividing his time equally between research and administration, but this had become difficult. He had to fight for funding. After pleading for UNESCO to find ways of finding more cash, he turned to his brethren in the Islamic countries, some of whom were awash in oil money, reminding them also of the history of the golden age of Islamic science. 'Be generous once again,' he beseeched. 'Create a talent fund.' His own personal contribution to this fund would be the $60 000 that he would receive in Stockholm on 10 December[32], which would go to help young students from Muslim lands. In keeping with the Ahmadi altruism, his share of the 1968 Atoms for Peace award had also gone to help young students. Soon afterwards, Salam passed through CERN, where he paid tribute to the Gargamelle team that had discovered the vital neutral current of the electroweak theory.

The Nobel was also an apotheosis for Sheldon Glashow and Steven Weinberg, who were sharing the prize with Salam for their 'contributions to the theory of the unified weak and electromagnetic interactions between elementary particles', and who had first met each other thirty years before at the Bronx High School of Science. Also at Stockholm in December 1979 was W. Arthur Lewis, who shared the Nobel Prize for Economics with Theodore Schultz. Born on the Caribbean island of St. Lucia in 1915, Lewis had become Professor of Economics at Manchester in 1948. Like Salam, he was a forerunner of the postwar multicultural Britain[33].

At the Nobel ceremony, Salam, a dashing figure in his traditional Punjabi dress, saw himself as an ambassador of the Third World. In the presentation speech, Bengt Nagel of the Royal Swedish Academy of Sciences explained the significance of electroweak unification, using an extract from John Updike's poem 'Cosmic Gall' about neutrinos:

'Neutrinos, they are very small.
They have no charge and have no mass
And do not interact at all.'

Nagel pointed out this was wrong – neutrinos do interact, through the weak nuclear force, which was now understood to be subtly related to the familiar phenomenon of electromagnetism. Had it not been for neutrinos, the achievements of Glashow, Salam and Weinberg would have been more difficult to verify. At the banquet that evening, speaking on behalf of the three Physics laureates, Salam thanked the Nobel Foundation and the Academy of Sciences before going on to add his personal gratitude in his native language, Urdu, and illustrating his own scientific motivation with a verse from the Holy Qur'an. He closed by recalling how Alfred Nobel had stipulated that his generosity would not discriminate against race or colour.

The Nobel award ceremony merits pomp and ceremony that only ancient monarchies, like Sweden, can muster. There were elaborate rehearsals for the ceremony, where Salam learned that he had to 'move forward when the trumpets sound, and stop when they stop.'[34] The traditional culmination of the festivities is the formal ceremony on 10 December, the anniversary of Alfred Nobel's death, but its proximity to the winter solstice means that it inherits Scandinavian winter traditions, where large amounts of alcohol attempt to compensate for the darkness. As a Muslim, Abdus Salam could not drink, but was used to attending events where those around him did. At Stockholm, Salam's laugh was drowned in raucous background. His children danced with Weinberg's.

The electroweak theory leans on W and Z messenger particles, respectively carrying the electrically charged and neutral components of the weak interaction. In 1979, all the evidence pointed to the existence of such particles, but they had never been seen. That would come four years later at CERN, and Carlo Rubbia and Simon van der Meer duly shared the 1984 Nobel for the achievement. In view of this missing link, the award of the prize to Glashow, Salam and Weinberg in 1979 could have been seen as premature. Leon Lederman (later to share the 1988 Physics Prize) commented wryly 'If the W and Z are not found, does this mean that Glashow, Salam and Weinberg will have to hand back their prize?'

Twenty years later, Gerard 't Hooft and Martinus Veltman duly shared the 1999 Nobel Physics award and enjoyed similar pomp for 'for elucidating the quantum structure of electroweak interactions in physics'. Had it not been for their independent work, the unification of electromagnetism and the weak force via the route charted by Glashow, Salam and Weinberg would have languished until experiments stumbled by chance across the neutral weak current. Their work also marked the renaissance of field theory, which in the 1970s emerged from the shadows to blossom in many areas of science, from elementary particle physics to cosmology. Throughout his valiant effort, Veltman had kept his nose so close to the grindstone that he had noticed little of what else was going on. For him, what Weinberg and Salam had done always remained a frog. Speaking at a physics history meeting in 1992, before Nobel recognition, the embittered theorist said 'if someone would have told me then [1971] about the 1979 Nobel Prize, I would have laughed.'[35] Veltman's book 'Facts and mysteries in elementary particle physics' (World Scientific, Singapore, 2003) has a photograph showing a confrontation between Steven Weinberg and the author, who says 'Weinberg and I do not see eye to eye on certain issues'.

At the Nobel Prize award ceremony at Stockholm in December 1979, the Salam contingent was about twenty strong, and many laureates were surprised to learn that it included two families.

Attending an antinuclear proliferation meeting in London in 1962, Salam had met Louise Johnson, then a physics undergraduate at University College London (UCL), who was helping with the meeting's administration. It was what the French call *un coup de foudre*, an emotional lightning strike, such as Salam had not experienced since seeing the inaccessible Urmilla at Government College, Lahore, some twenty years before. After graduating from UCL, Louise Johnson began research at the Royal Institution, London, on the structure and properties of large biological molecules. The prime research tool for this work is X-rays, whose tiny wavelengths can probe the deep interior of biomolecules and reveal their structure.

If X-ray crystallography had become a jewel in the British scientific crown, the Royal Institution was almost a crown in its own right. Through the centuries it has astutely identified important new research trends and nimbly contributed to milestone discoveries. It was there in the early nineteenth century that Humphry Davy made his showman demonstrations of the power of electrolysis, discovering new elements such as sodium and potassium. One who saw Davy's science spectaculars was Michael Faraday, who went on to inherit Davy's professorial chair at the Royal Institution, shifting its course from electrochemistry to the new science of electromagnetism. The Royal Institution's research direction changed again in 1923 when William Henry Bragg was appointed its Director. He had opened up the use of X-rays to 'illuminate' the deep interior of atoms and molecules, and his son, William Lawrence Bragg, went on to use X-rays to uncover molecular structure. Father and son had shared the 1915 Nobel Physics Prize for their work.

This new research direction in British science was emphasized in 1938 when W. L. Bragg became head of the Cavendish Laboratory at Cambridge, as successor to Ernest Rutherford. In this prescient move, the Cavendish turned away from Rutherford's tradition of nuclear research and set out on a fresh route. This carefully orchestrated research effort gradually built up to a crescendo, leading to the unravelling of the molecular structure of DNA by Francis Crick and James Watson in 1953. In that year, W. L. Bragg (now Sir Lawrence) moved from the Cavendish to the Royal Institution, continuing the X-ray analysis tradition and helping to build the new field of protein crystallography. In 1965, David Phillips, working under Bragg, built automated instrumentation to elucidate the structure of enzymes, the natural catalysts in living cells that bring about chemical reactions essential to life. Phillips' subject was lysozyme, found in egg white and in human tears and saliva, and that acts as an antibacterial agent. This was Louise Johnson's introduction to molecular biophysics research. Her studies of the enzyme's substrate binding began to probe the mystery of how it was able to break down the walls of bacterial cells. After completing her doctorate at the Royal Institution, she spent a year at Yale before moving to Oxford in 1967, where David Phillips had meanwhile become Professor of Molecular Biophysics.

Salam and Louise Johnson were married in a Muslim wedding in London in 1968. An unlikely witness was Paul Matthews, Salam's long-time research partner and professor at Imperial[36]. In Islamic terms, his new relationship was a marriage, so Salam was following the edicts of a religion that expressly forbids fornication[37], but on the other hand it was sufficiently distant from a union that had taken place between cousins in Pakistan as not to cause alarm. The freedom and support that Salam's unorthodox lifestyle required was freely given on all sides, and the unconventional arrangement worked. By deft planning and attention to detail, and by supreme forbearance by those involved, Salam was able to manage his unconventional matrimonial affairs, shuttling between Trieste, London and Oxford. Salam was discreet about all this, but on the other hand did not keep it secret. His 'second family' became regular summer visitors at Trieste.

Louise Johnson went on to take Oxford's David Phillips professorial chair of Molecular Biophysics, and in 1990 was elected a Fellow of the Royal Society. In 2003, she was honoured with the title of Dame of the Order of the British Empire (DBE) for her services to science. As a Briton, she assumed the title 'Dame Louise': in 1989, Abdus Salam, as a foreigner, had not been able to take the title 'Sir' when awarded his Knighthood of the Order of the British Empire (KBE). In any case the title 'Sir Abdus' would have resurrected the incongruity of sundering the essentially dual Arabic name of Abdus Salam, which by then most people, including Salam himself, had grown used to.

Salam was particularly pleased that the age difference between his two sons, Ahmad, by Amtul Hafeez, born in 1960, and Umar, by Louise Johnson, born in 1974, was the same as that between his father, Muhammad Hussain, and his uncle and father-in-law, Ghulam Hussain. Salam knew how close and influential those brotherly ties had been.

REFERENCES

1. Such as the isotopic spin that differentiates a proton from a neutron in nuclear terms
2. Crease, R., Mann, C., *The second creation, makers of the revolution in twentieth-century physics* (New York, Macmillan, 1986)
3. http://www.hull.ac.uk/php/masrs/reminiscences.html
4. CERN, Pauli archives
5. Salam, A., *A life of physics*, in Cerderia, H. A., Lundqvist, S. O., (ed.) *Frontiers of physics, high technology and mathematics* (Singapore, World Scientific, 1990)
6. Salam, A., *A life of physics*, (1990)
7. Goldstone, J., Salam, A., Weinberg, S. *Broken symmetries*, Physical Review **127**, 965–70 (1962).
8. Crease, R., Mann, C., (1986)
9. Weinberg, S., *Tracing the mechanism of electroweak symmetry breaking*, in Cerderia, H. A., Lundqvist, S. O., (ed.) *Frontiers of physics, high technology and mathematics* (Singapore, World Scientific, 1990)
10. Salam, A., Ward, J. C., *Weak and electromagnetic interactions*, Nuovo Cimento **11**, 568–77 (1959)
11. http://www.opticsjournal.com/JCWard.pdf
12. Salam, A., Ward, J. C., *On a gauge theory of elementary interactions*, Nuovo Cimento, **19**, 165–70 (1961)
13. Crease, R., Mann, C., (1986)
14. The possibility had been looked at in the 1930s, notably by Kemmer
15. http://www.opticsjournal.com/JCWard.pdf
16. Salam, A., Ward, J. C., *Electromagnetic and weak interactions*, Physics Letters, **13**, 168–71 (1964)
17. Jackson, J. D., *Snapshots of a physicist's life*, Annual Reviews of Nuclear and Particle Physics, **49.1**, 1–33 (1999)
18. Brout, R., *Notes on spontaneously broken symmetry*, in *The rise of the standard model*, Brown, L., Dresden, M., Hoddeson, L., (ed.) (Cambridge, CUP, 1997)
19. http://olympus.het.brown.edu/chep/stlouis-v3.pdf.
20. Weinberg, S., *The red camaro*, in *One hundred reasons to be a scientist*, (Trieste ICTP, 2004)
21. Weinberg, S., *A model of leptons*, Physical Review Letters, **19**, 1264–66 (1967)
22. Salam, A., *Weak and electromagnetic interactions*, in *Elementary particle theory*, Svartholm, N., (ed.) (Stockholm, Almqvist & Wiksell, 1968)
23. Perkins, D., *The discovery of neutral currents*, in Brown, L., Dresden, M., Hoddeson, L., (ed.) *The rise of the standard model*, (Cambridge, CUP, 1997)
24. Salam, A., Komar, A., *Renomalization problem for vector meson theories*, Nuclear Physics, **21**, 624–30 (1960) and Salam. A., *Renormalizability of gauge theories*, Physical Review, **127**, 331–4, (1962)

25. Veltman, M., *The path to renormalizability* in *The rise of the standard model*, Brown, L., Dresden, M., Hoddeson, L., (ed.) (Cambridge, CUP, 1997)
26. Veltman, M., *The path to renormalizability* (1997)
27. Veltman, M., *The path to renormalizability* (1997)
28. Coleman, S., *The 1979 Nobel Prize in Physics*, Science **206**, 1290–2 (1979)
29. http://nobelprize.org/physics/nomination/index.html
30. Smyth had authored the influential report 'Atomic Energy for Military Purposes', published by the US Government in August 1945, immediately after the attacks on Hiroshima and Nakasaki, and that explained the science of the bombs. (Fission is nuclear, rather than atomic energy, but this report preferred to promote a name that, although technically incorrect, was then more familiar to the public. The weapons that exploded at the end of the Second World War should have been called 'nuclear', rather than 'atomic' bombs, and the Geneva conferences should really have been called 'Nuclei for Peace'.)
31. http://www.opticsjournal.com/jcward.pdf
32. Salam A., Address to the UNESCO Executive Board, in *Ideals and realities*, Hassan, Z., Lai, C. H. (ed.) (Singapore, World Scientific, 1984)
33. de Greiff, A., *Abdus Salam, A migrant scientist in post-imperial times*, Economic and Political Weekly, 21 January 2006, 228.
34. Ghani, A., *Abdus Salam: A Nobel Laureate from a muslim country*, (Karachi, published privately, printed Ma'aref, 1982)
35. Veltman, M., *The path to renormalizability* (1997)
36. Salam would have preferred two Muslim witnesses to his new marriage, and this was duly rectified in a second ceremony in 1973.
37. Sura 4 of the Holy Qur'an offers much matrimonial guidance. After Verse 3 points out that a man can have several wives, Verse 24 says 'Seek their hands by means of your wealth, marrying them properly and not committing fornication to pursue your lust'.

13

Quark Liberation Front

Throughout his research, Salam sought for patterns, similarities in behaviour between subnuclear particles that were superficially very dissimilar. His approach contrasted with that of many of his contemporaries, who viewed categories of particles essentially as items on a menu where the particle masses were marked like prices, with the lightest ones at the top of each category and therefore the most familiar, and the heaviest, of interest only to connoisseurs, relegated to obscurity at the bottom. Salam held that such an implicit classification in terms of mass clouded the real physical meaning: for him, mass, despite being the easiest particle attribute to comprehend, was almost a detail[1].

Clinging to this viewpoint, he still had one outstanding assignation with his lifelong quest to unify the forces of Nature. He had left it hanging in the final sentence of his talk at the 1968 Nobel Symposium in Gothenburg. This electroweak unification had lashed together two apparently very different limbs of Nature, originally thought to act independently of each other – the electromagnetic force that holds atoms together, and the weak nuclear force that can make the nuclei of those atoms fall apart. The picture Salam had painted at Gothenburg used only those particles, like electrons and neutrinos, which interact through the weak force. This facet of Nature had first been recognized in beta decay, in which neutrons and protons exchange roles. However, protons and neutrons were now understood to be made up of Gell-Mann's quarks. Locked inside nuclear particles, quarks are difficult to perceive, but the furtive weak force can still seek them out and refashion them, transforming the quark composition of a nuclear particle. Salam's next step in unification had to include quarks.

Quarks intrigued and puzzled people. At a simple level, they explained the luxuriance of the subnuclear landscape. But at a less simple level, at the beginning of the 1970s nobody could understand how quarks could actually stick together and form so much exotica. Pauli's

Exclusion Principle, an inviolate quantum law, said that when such particles (with half-integer spin) cohabited, each one had to have entirely different quantum labels. Each valid quantum slot was individually numbered and could accommodate just one participant. There was no room for more. However, some of the required quark combinations did not satisfy this decree. How could quarks flout these rules and still stick together?

This was not the only quark eccentricity. In addition, they were supposed to carry their electricity as unfamiliar fractional charges, either one-third or two-thirds that of the electron, the classic quantum of electricity. Fractional charges were so unfamiliar as to be almost meaningless, like Harry Potter being instructed to proceed to platform 9¾. Most physicists closed their eyes to these problems, and peppered their papers with quarks. When questioned, they hedged their bets and would not admit that such iconoclastic particles actually existed deep inside protons and neutrons. Quarks, they said, were merely abstractions: a sort of subnuclear bank statement, invaluable for calculations, but existing only on paper.

Gell-Mann had labelled his quarks with flamboyant disregard for intellectual dignity: 'up' and 'down' making up everyday protons and neutrons, with a third 'strange' quark for more exotic nucleons. To overcome the problems with the Pauli rule, one hesitant suggestion was to endow quarks with an additional charge-like attribute. For this, the threefold SU(3) symmetry developed by Gell-Mann and Ne'eman was peeled apart to give two separate SU(3)s. The first covered the up/down/strange quark triplet, while the second reflected a new triad of quark labels that acquired the name 'colour' (apparently anything to do with quarks had to be given a banal name). With this additional flexibility, the quantum labels on three quarks clustered together could be arranged to respect the Pauli rule. The meaning of these new colour attributes soon became clear – they were the three different 'charges' seen by the interquark force, analogous to the positive/negative duality of electromagnetism, where unlike charges attract and like charges repel. Quarks stick together in such a way that the net colour charge is always zero ('colour' coming from a crude analogy with the three primary colours, red, green and blue, which combine to yield colourless 'white' light). This new colour label could also be used to generate an extra component of electric charge, supplementing Gell-Mann's fractional charges and endowing quarks with conventional electricity.

In 1972, this idea was still on the market, and appealed to those who would still not buy quarks with fractional charges. One was Salam.

He had now acquired a new research partner. Born in Orissa, India, in 1937, Jogesh Pati was a theoretical physicist working at the University of Maryland. During a sabbatical in 1972, he returned to his home country for six months at the University of Delhi. A sabbatical period is meant as an opportunity to recharge intellectual batteries and work on research ideas that have been pushed into the background because of other commitments. Instead, Pati quickly experienced the sort of intellectual isolation that Salam had felt in Lahore twenty years earlier. In 1971, theoretical physicists had been startled by the news of 't Hooft and Veltman's revelations, and field theory had zoomed back into fashion. But Pati, isolated in Delhi, had not known until he stopped over in Trieste before returning to the US. Pati hoped he was not too late to join the party. Over tea after a Trieste seminar, he suggested to Salam that there should be some deep reason why the proton and the electron are so different yet contrive to carry equal but opposite amounts of electricity. Protons contain quarks, but the electroweak theory worried only about electrons and neutrinos, and said nothing about quarks. Bringing all Nature's ingredients together in some new symmetry scheme might reveal a reason for the contrariety of these particles and the forces they feel. Salam leapt at the suggestion: 'That seems like an excellent idea! Let us develop it immediately.'[2]

Thus began a collaboration between two scientists from the Indian subcontinent who, for different reasons, had chosen to work elsewhere in the world. Pati explains that because of their upbringing, the elder Salam was automatically assumed to be the wiser, but Pati was still allowed to be politely critical. He cites a typical exchange after a mutual disagreement:

Salam: 'My dear Sir, what do you want, blood?'
Pati: 'No, Professor Salam, I would like something better'.

Pati and Salam's bold idea was to supplement the three quark colours with a fourth label, this time assigned not to quarks but rather to particles like electrons and neutrinos, which felt only the weak force. With four colours, the full symmetry was enlarged to a dual SU(4). Having declared their imaginative symmetry, Salam promptly tried to shackle it by giving their quarks integral charges, leading to some startling new predictions. However, at Pati's insistence the fractional charge route was not ruled out, and their theory always offered a choice of quark charge.

With additional mechanisms now acting on both quarks and weakly interacting particles, quarks could transform into electrons or neutrinos. If quarks could thus vanish, then so could protons, the bedrock particle of the whole nuclear world. With transient protons, no atom could be eternal: the whole Universe had a shelf life. This was not the aim of their theory. It was Pati that initially realized they had an unexpected child that Salam was initially reluctant to accept, and wanted to reconstruct their theory to abort it.

Their summer 1972 brainstorm was interrupted when Pati had to return to Maryland, and Salam became caught up in his habitual hectic travel schedule. Separated by an ocean and hamstrung by Salam's itinerary, the collaboration continued at a more leisurely rate, with Pati as its initial spokesman. Their new theory arrived at that year's late summer conferences too late to be presented in its proper context. Its public debut as a brief mention in a summary talk attracted little attention.

It took Salam some time to get used to the idea of an unstable proton. However, it was not the first time that such supreme subnuclear indignity had been suggested. In 1967, the Soviet physicist Andrei Sakharov had conjectured it to explain why a Big Bang, which should have wrenched exactly equal amounts of matter and antimatter from the primordial void, could have mutated instead into a skewed Universe in which all natural antimatter seems to have disappeared. The Big Bang's balance of particles and antiparticles has been replaced by the familiar but clumsy electrical pairing of protons and electrons. Fortunately, any required proton instability was miniscule, even in the grandiose accounting of cosmology: compared to the age of the Universe (about 14 billion years), the average lifetime of a proton in years should be larger than the age of the Universe – expressed in nanoseconds. But there are enough protons in a gram of hydrogen to reconcile such discordant numbers. It was not a fatal recipe: the arithmetic said that a person could live for a hundred years before having a good chance that just one corporeal proton would disappear. But by monitoring thousands of tons of protons, perhaps the tell-tale signature of proton decay could be seen.

Pati and Salam had unlocked a door to an unfamiliar world, but a map marked with integer-charge quarks at least made it less difficult to navigate. Quarks with integer charges should also be visible, not hide themselves inside protons and neutrons, and Salam started going around wearing a 'Quark Liberation Front' badge on his jacket lapel, above a top pocket bulging with spectacles and pens. Quark instability was an attractive explanation for why quarks had never been seen – if they were let loose from their nuclear prisons, they would promptly decay. As the Italian theorist Daniele Amati said later of this idea: 'the price of liberty is death'[3].

After Salam had eventually reconciled himself to unstable protons, the theory was published in the summer of 1973[4]. Pati and Salam now felt that the implications were too important to be buried in such a long paper. Their ancillary proposal 'Is Baryon Number Conserved?' (physics-speak for 'Is the proton unstable?') was completed in Trieste in the early summer of 1973, just a few days before Pati was due to return to the US for a six-week residence at Brookhaven National Laboratory, New York. Pati explains[5] that Salam wanted to submit the paper to *Physics Letters* in Europe so as to avoid being delayed by a lengthy US refereeing procedure. Pati on the other hand favoured the American *Physical Review Letters*, and as he was going to Brookhaven, he would personally take it to the Chief Editor, Samuel Goudsmit. After doing so, Pati gave a talk at Brookhaven that aroused the interest of senior scientist Maurice Goldhaber, whose distinguished career had started with Rutherford in Cambridge. Although the discussion at Brookhaven raised Pati's hopes, the paper was soon turned down by *Physical Review Letters*. The disappointed Pati invaded Goudsmit's office. Prepared to wedge his foot in a closing door, he was instead surprised to get a warm reception from Goudsmit, who said that he had recently met Salam in Europe. With Salam applying unrelenting pressure from afar, Goudsmit was now sandwiched by the two authors. Within a few hours, the referee's decision was overruled and the paper accepted[6].

Salam had meanwhile rushed off from Trieste to attend the major 1973 particle physics meeting in Aix-en-Provence, France. At that meeting, physicists learned that the neutral current effects predicted by electroweak unification had at last been found. The jubilant Salam was invited to add a few words after Steven Weinberg's talk. Instead of gloating over the success of their electroweak theory, he was instead so excited by the implications of his new unification scheme that he urged experiments to start looking immediately for proton decay. Instead of a syrupy electroweak eulogy, the audience was served a dish of blatant speculation, too spicy for their liking. Few realized the implications of what Salam was saying.

To search for proton decay, the trick was to find a sample of protons big enough to amplify the tiny signal (why not try the Moon? suggested an unfazed Salam later[7]). Soon after his talk at Aix-en-Provence, Salam set about motivating new experiments to look for the effect. He had seen how Weinberg, after the revival of interest in the electroweak idea in 1971, had pushed experiments to look for the missing neutral currents. One place where proton decay could already have been seen without anybody realizing was the big experiments studying rare neutrino effects. Neutrino pioneer Fred Reines took Pati and Salam to see William Wallenmeyer, the official treasurer for particle physics in the US Department of Energy. Construction work soon began for a huge experiment, led by Reines and by Maurice Goldhaber, to look for signs of proton decay. Arrays of photosensitive detectors would line an 8000-tonne tank of water 600 metres underground in the Morton-Thiokol salt mine at Fairport, Ohio. At a meeting of the UN Advisory Committee on Science and Technology in New York, Salam again bumped into Mambillikalathil 'Goka' Menon, who had already been influential in his international plans. Listening to Salam's proton-decay plea, the Indian scientist quickly calculated that a hundred-ton detector mounted deep underground to screen the cosmic debris of outer space might have a chance of seeing something. A hundred tons of steel is not that big – a small roomful. Mounted two thousand metres underground in an Indian gold mine, it became another eye scanning for the ultimate instability of nuclear matter.

But before these new experiments even began working, in November 1974 the world of particle physics was shaken by a totally unexpected discovery. Two different experiments, one at the Stanford Linear Accelerator Center (SLAC), California, and another at Brookhaven, found the huge signal of a new and remarkably stable subnuclear particle in a previously unexplored energy range. With the two experiments unaware of each other's discovery, it was called psi by Burton Richter's team at SLAC and J by Sam Ting at Brookhaven[8]. Every available theory was pressed into service to try to understand the mysterious subnuclear newcomer. Pati and Salam's was one. Their paper 'Lepton Number as the Fourth Color' which had appeared earlier in 1974[9] contained a whole list of predictions. The new signal, said Salam, was on that list. With his fixation on integer-charged quarks, he was adamant that the new effects were due to 'colour gluons' that carry the interquark forces. Instead, Richter and Ting's J/psi was soon found to be due to a fourth quark, 'charm', which augmented the up/down/strange triplet, extending the subnuclear map in a new direction.

Before this dust had even settled, Salam proposed holding a major international meeting in Trieste for the following summer. 'Phenomenology in High Energy Physics and the Missing Particles' was physics-speak for the inexplicable new results and whatever would follow in the coming months. All the major players in the game were invited. But Salam's UNESCO-funded institute suddenly found itself cold-shouldered, boycotted by prominent scientists in the US-led backlash to UNESCO's political tub-thumping. With prominent US laboratories not allowing their experimental results to be displayed at Trieste, Luciano Bertocchi was despatched across the Atlantic as an ambassador. But as long as Trieste was UNESCO-funded, the disapprobatory scientists maintained, they would not attend[10]. With a shadow falling over both his theory and his institute, Salam felt rejected. More rebuffs were to come.

In his work with Pati, Salam had insisted on integrally charged quarks from the outset. Soon after the idea was launched, new experiments comparing the effects of electron and neutrino beams showed that quarks indeed carry fractional charges, and the model, in the form that Salam would have liked it, looked to have missed the boat. But Pati's dogged resistance to Salam's preconception meant that one leg of the theory could survive, and its imaginative approach still continues to influence thinking. Their pioneer work had already pointed the way for others, once the vital role of colour in the interquark force had been fully understood. Howard Georgi and Sheldon Glashow at Harvard made another foray into unifying quark and electroweak effects. Georgi said 'I decided to look at the Pati–Salam model.... I went through the exercise of seeing how it worked with the colour [symmetry] unbroken.... ...I realized what a beautiful idea the Pati–Salam [symmetry] is'.[11] The 1974 Georgi–Glashow model – *The unity of all elementary particles*[12] was not correct either, but it included fractionally charged quarks and is now presented in textbooks as the prototype of a grand unification of electroweak and quark forces.

As well as inspiring other theoretical ideas, the Pati–Salam unification scheme was eventually to have an unexpected windfall. They had catalysed the development of a new generation of huge experiments to search for signs of proton decay and related effects. While none were found, these huge eyes on the cosmos were open on 23 February 1987, when a bright light suddenly appeared in the sky. Astronomers were witnessing a supernova, the death throes of a large star as it ran out of nuclear fuel and collapsed, releasing a tide of stellar debris of which the visible light is only the envelope. Big astronomical telescopes see such things in the distant heavens about once a month, but supernovae bright enough to be seen with the naked eye only happen once every few hundred years. This time, the big underground detectors caught the particulate wash of a supernova signal. It was the birth of a new science – particle astronomy.

His work with Pati had been Salam's second attempt to launch an ambitious theory of elementary particle physics from Trieste. The collaboration, which lasted from 1973 to 1984, was also Salam's most prolific period in terms of scientific papers produced. Before this, the most papers he had written in a single year had been ten, in 1961. In 1973 there were 13, in 1974 15, in 1975 18, in both 1976 and 1977 ten, and in 1978 13, many of these written in collaboration with Pati.

For their 'grand' electronuclear unification, Pati and Salam had set out to combine quarks and weakly interacting particles, elements of the Universe that previously had been considered totally distinct from each other. It had pointed a way forward: quantum rules once thought inviolate could instead be mere convention. Proton decay, once inconceivable, is still a major objective, and experiments continue to search diligently. To separate principle from preconception, Pati and Salam had shown that customary quantum rules needed to be re-examined.

In another apparently immutable binary classification, subnuclear particles followed separate religions according to whether their spin was an integer or a half-integer quantum multiple. Integer spin particles ('bosons', after Satyendra Nath Bose) have no quantum-energy restrictions, and any number can sit happily in whatever quantum slot is allowed. Half-integer spin particles ('fermions', after Enrico Fermi) obey Pauli's exclusion rule – only one particle at a time can sit in an allowed quantum slot. Bosons transmit interactions between matter made up of fermions. Why did fermions and bosons look so different? It was redolent of the proton–electron duality that had initially intrigued Salam and Pati.

In the continual push to reconcile the enormity of Einsteinian relativity with the minutiae of the quantum world, quantum gravity was being revisited. In 1970, Salam, Christopher Isham and John Strathdee had shown how infinities in quantum electrodynamics could be suppressed by subtle gravity-controlled effects. In human experience, the quantum world and gravity are poles apart, but under less familiar conditions, maybe an unexpected interplay could be revealed. Certainly gravity and the quantum world would approach under conditions involving other forms of substance – microentities as compact as a proton but as 'heavy' as a grain of dust, and that could actually be weighed on sensitive mechanical scales. With such ponderous particles, the gravitational effects that are invisible in the microscopic domain would become comparable to subnuclear forces, and the physics landscape focused into a small more compact frame.

(Describing his collaboration with Salam, Isham recalls the monk-like devotion to duty that this intense work required, and the persistent optimism that characterized Salam's approach. He cites varying degrees of Salam's personal scheme of mathematical induction: if something is true several times, then perhaps it is always true; if something happens once, then maybe it is always true; and if Salam wanted something to be true, then perhaps this could be arranged. A physics paper could be drafted with an introductory premise and its final implications, and Salam's research lieutenants would be left to link the two during one of his frequent trips. While such an approach was often embarrassingly optimistic, sometimes Salam's intuition would amaze his collaborators. When they asked him how this came about, he would solemnly point his finger to the sky.)

Fresh ideas suggested that unfamiliar superheavy entities might not be infinitesimally small points with zero dimensions. Instead they could be one-dimensional strings that vibrated in some invisible inner space. Different particles corresponded to the various harmonics of these strings, in the same way that a violin string sounds different whether it is bowed or plucked. An arpeggio of possible vibrations could include several particles, both bosonic and fermionic. In such theories, the conventional distinction between matter particles and interaction particles disappeared: matter particles could be bosons and messenger particles could be fermions. A gust of wind suddenly scattered ideas in new directions. In 1973 Julius Wess and Bruno Zumino[13] put aside the string context that had spawned the ideas and pointed out the far-reaching implications of new ways of relating subnuclear particles with symmetry schemes involving 'supergauge' transformations.

Smelling the scent of breakthrough, Salam and his longtime Trieste collaborator John Strathdee were quickly on the scene, even before the Wess–Zumino ideas had been published. Strathdee did not like the term 'supergauge' for a symmetry that was rigid and 'global'. In a footnote to the first paragraph of one of their first papers on the new development[14], Salam and Strathdee pointed to the pedigree of the term, adding 'since the word "gauge" has come to be associated more commonly with "gauges of the second kind", or local symmetries, it is confusing to use supergauge to describe what is indeed a global symmetry of fermions and bosons. We suggest therefore that the expression "supersymmetry"[15] might be more appropriate for the global concept and reserve the word "gauge" for local symmetries'. Their suggestion stuck and a new word entered the scientific vocabulary[16].

In supersymmetric theories, fermion and boson contributions can enter the calculations with opposite sign, providing a useful way of cancelling out potentially infinite contributions. It was an elegant quantum framework for new theories of gravity – 'supergravity'. The mathematical skill of John Strathdee, emerging from under the shadow of Salam, played an increasingly important role, particularly fruitful in this work during the late 1970s and early 1980s, which constructed techniques for handling 'superfields'. In the 1980s, supersymmetry and 'superstrings' became fashionable and developed into a major theoretical industry, providing fertile ground for the skills of talented young theorists all over the world. A new stage had been constructed for the ultimate theory that encompassed all of Nature's agents. Salam was a player on this stage, but was not alone in becoming increasingly concerned by the conceptual clash between the four dimensions of space-time and the ten of multidimensional string theory, even though Einsteinian gravity can be made to crystallize from it.

Salam's science was driven by prolific collaborations, fuelled by diverse energies and talents. His joint efforts with Paul Matthews lasted until the mid-1960s and dovetailed smoothly into the Trieste-based teamwork with Robert Delbourgo and John Strathdee. In the 1970s a new symbiosis opened with his work on electronuclear synthesis with Jogesh Pati. From the early 1980s, a new collaborator at Trieste was Seifallah Randjbar-Daemi. Born in Tabriz, Iran, in 1950, Randjbar-Daemi first met Salam as a student at Imperial College in the mid-1970s, where he completed his PhD with Tom Kibble. After moving to Trieste in 1980, Randjbar-Daemi increasingly worked with Salam, often together with John Strathdee, on matters of the moment, frequently on new directions for fundamental theories. Randjbar-Daemi and Salam did their science in English, but occasionally Salam would entertain his fellow-physicist by reciting Farsi poetry in a Pakistani accent. Their research was conducted by Salam in typical style, with Randjbar-Daemi and Strathdee at the blackboard while Salam frequently kept one ear clamped to the telephone, his utterances alternating between abstract suggestions to his research colleagues in the room and on more tangible matters into the telephone.

For some twenty years after the founding of the Trieste centre, Salam had automatically remained the leader of its elementary particle theory group. As the centre's scientific interests spread, new groups came into being, demanding an official corporate structure, with key responsibilities clearly designated. But particle theory and Salam's work always had a special role as the centre's core business. There was no formal head of particle theory. Salam ran it, as the Sun rose daily in the sky, in addition to his duties as the Institute's Director. Salam's particle theory was Trieste's particle theory. But in the mid-1980s, with Salam preparing his bid for UNESCO leadership, it became clear that Trieste's management structure needed to be more clearly defined, and Randjbar-Daemi was formally appointed as head of the elementary particle theory group. His understanding had been that no administrative work would be required, with his responsibilities limited to overseeing scientific research. However, after unexpectedly finding himself part of the organization behind Trieste's spring and summer schools, he complained. Salam's immediate response was that these were not mere administrative duties but an integral part of the scientific effort, adding, in his characteristic way, 'if you don't do it, whom else can I ask'. Salam still ruled over his centre, but the fact that someone from South Asia had formally taken over responsibility for the flagship particle theory group must have given him pleasure.

In the 1990s, Salam turned to what would be his final contributions to science. The flurry of discoveries and the scramble for new theories that had so characterized subnuclear science in the 1950s and 1960s had come to an end. Idle speculation on superstring theory had taken its place. Physically he was weaker, but his imagination was still bright, and needed a fresh focus. The discovery in the mid-1980s of a new class of 'high-temperature superconductors' had stimulated efforts to understand this mysterious new effect, just as in the 1950s quantum theory had been harnessed to the investigation of classical superconductivity at liquid helium temperatures. With Randjbar-Daemi and Strathdee, Salam returned to a field of physics he had last visited in 1954, while still at Government College, Lahore.

After dabbling with superconductivity, Salam looked in another new direction. His 1956 theory of a left-handed neutrino had shown him the importance of left–right sensitivity. In 1848, Louis Pasteur had discovered how certain molecules could twist the axis of vibration of light waves to the right or to the left, and had suggested that the wider cosmos too should contain such directional pointers. Having shown the need for left-handed neutrinos, C. N. Yang suggested in his 1957 Nobel lecture that molecules too could emerge left-handed and lead to wider effects. Compelling evidence came in 1953 when Francis Crick and James Watson showed how organic molecules have an inbuilt direction. The hereditary template of DNA molecules that control the continuity of life is arranged in a twisted spiral. Life is permeated by subtle left–right distinctions: our bodies are built of left-handed amino acids and right-handed sugars, but unusual oppositely handed molecules also exist, and can trigger disease or immunological effects[17]. The meteorite that fell near Murchison, Australia, in 1969 was found to contain left-handed amino-acids, showing that these are not some terrestrial accident.

After initial work in collaboration with John Strathdee and Louise Johnson, in 1991 Salam suggested in a broad review paper[18] how a tiny left–right sensitivity in nuclear physics could have driven the evolution of left–right asymmetries in organic molecules. The text, dictated to a Trieste typist, included a meandering series of technical appendices. While nuclear effects can generate a left–right sensitivity, the effect is far too small to induce the levels seen in terrestrial life, unless they become amplified. In a later, more technical paper, Salam used detailed arguments to suggest how this could have happened[19]. He was not the first to consider such possibilities. After Pasteur, left–right asymmetry had been periodically revisited. Pierre Curie had dabbled with it in 1894. With the underlying science still in its infancy, the colourful British geneticist J. B. S. Haldane alluded to it as he boldly carved out his 1932 opus *The causes of evolution*. The British biochemist Leslie Orgel revisited the subject in another milestone book in 1973[20]. With the origin of left–right asymmetry on firmer ground, Haldane returned to it after parity violation had been discovered in the mid-1950s[21]. Murray Gell-Mann's thinking on the subject was reflected in his book *The quark and the jaguar*[22]. Salam apart, the consensus was that a link between atomic physics and molecular biology was too tenuous and should instead be attributed to a simple accident of evolution, in the same way that the outcome of a football match is biased by the initial toss of a coin to decide which team kicks the ball first. Salam's forays into these realms were disapproved of by most of his Trieste colleagues. Although he acknowledged help from many sides, notably Trieste astrobiologist Julian Chela-Flores, his name appears alone. Apart from invited review articles and conference talks, these were the first papers signed by him alone for some ten years. They were also the last.

REFERENCES

1. Salam, A., Why they gave me the Nobel Prize in Physics, talk given at Oslo, 13 December 1980, available via ICTP intranet, Trieste.
2. Pati, J. C., *Twenty years later* in *Salamfestschrift*, Ali, A., Ellis, J., Randjbar-Daemi, S., (ed.) (Singapore. World Scientific, 1994)
3. Quoted in Pati, J. C., Salam, A., *Lepton number as the fourth color*, Physical Review **D10**, 275–89 (1974)
4. Pati, J. C., Salam, A., *Unified lepton-hadron symmetry*, Physical Review **D8** 1240–51, (1973)
5. Pati, J. C., private communication, February 2007
6. Pati, J. C., Salam, A., Physical Review Letters, **31**, 661 (1973)
7. Pati, J. C., Salam, A., B Sreekantan, International Journal of Modern Physics A, **1**, 147 (1986)
8. In 1976 Richter and Ting went on to share the Nobel Physics prize for their discovery, now known as the 'J/psi' particle to commemorate its simultaneous discovery.
9. Pati, J. C., Salam, A., Physical Review, **D10** 275–89 (1974)
10. Trieste archives, G118
11. Georgi, H., *Grand unified theories* in *history of original ideas and basic discoveries in particle physics*, in Newman, H., and Ypsilantis, T., (ed.) *History of original ideas and basic discoveries in particle physics*, (New York, NY, Plenum, 1996)
12. Georgi, H., Glashow, S.,*Unity of all elementary particles*, Physical Review Letters, **32**, 438–41 1974
13. The ideas were independently suggested in the Soviet Union, but this was not immediately realized elsewhere.
14. Salam, A., Strathdee, J., *Supersymmetry and non-Abelian groups*, Physics Letters **51B** 353–55 (1974)
15. Actually the paper used "super-symmetry" with a hyphen

16. Confusingly, Pati and Salam had used it in another context in 1973 to describe their new unification theory
17. McManus, C., *Right hand, left hand*, (London, Weidenfeld & Nicholson, 1992)
18. Salam, A., *The role of chirality in the origin of life*, Journal of Molecular Evolution, **33**, 105–13, 1991
19. Salam, A., *Chirality, phase transitions and their induction in amino acids*, Physics Letters, **B288**, 153–60 (1992)
20. Haldane, J. B. S., *The causes of evolution*, (London, Longmans, 1932)
    Orgel L., E., *The origins of life, molecules and natural selection*, (London, Chapman and Hall, 1973)
21. McManus (1992)
22. Gell-Mann, M.,*The quark and the jaguar*, (New York, NY, Little, Brown, 1993)

14

Demise

Salam was in favour in Pakistan when President Muhamad Ayub Khan remained in control. But beneath the surface, the tide was beginning to turn. Minister of Fuel and Power in Ayub Khan's government from 1958 was Zulfikar Ali Bhutto, a sophisticated politician from a landed Sindh family, educated at the University of California at Berkeley, and at Oxford, and who had qualified as a barrister in London. Because of his portfolio, Bhutto's involvement in the ongoing nuclear effort in Pakistan began to overlap with the formative plans of Salam, still the President's science advisor, and of Ishrat Hussain Usmani, who had taken over as head of Pakistan's Atomic Energy Commission. While Salam and Usmani concentrated on peaceful uses of nuclear energy and technology, Bhutto had other designs.

After India's defeat in 1962 in a border clash with China, followed by China's first nuclear detonation in 1964, influential voices in India began to clamour for nuclear weapons. The head of the Indian Atomic Energy Commission was Homi Bhabha, who thirty years before had been one of the first scientists to calculate quantum effects in electrodynamics. He died in 1966 *en route* to CERN when his plane crashed into Mont Blanc, but he had already set wheels in motion that would ultimately lead to the first Indian nuclear detonation in May 1974. Pakistan now saw a mushroom-shaped cloud as well as the traditional mistrust of India. In the lead-up to the nuclear test, Bhutto said 'If India builds the bomb, we will eat grass or leaves, even go hungry, but we will get one of our own, we have no alternative'.[1] Dismissed from Ayub's government in 1967 after having become Foreign Minister and playing a key role in an abortive scheme that led to the 1965 war with India over Kashmir, Bhutto went his own way and set up the Pakistan People's Party. In 1968 Ayub Khan was deposed in the face of popular pressure and replaced by General Yahya Khan, who among other things had less interest in science and technology. Salam's influence on Pakistani affairs began to wane. However, during this period, Salam exhorted the President to remedy Pakistan's isolation from world science, and convinced Finance Secretary M. M. Ahmad, passing through London, to earmark 50 lakhs (5 million) rupees to fund a Pakistan Science Foundation[2].

After more than a decade of military rule, Yahya Khan declared Pakistan's first open general election in 1970, a move of which he was proud. 1570 candidates from 25 parties contested some 300 seats. Ironically, these free elections set the scene for fresh chaos that was to rock the fragile nation that had emerged in 1947. Bhutto's Pakistan People's Party secured 81 of the 138 parliamentary seats for West Pakistan, mainly in Sindh and the Punjab, while Sheikh Mujib's Awami league won almost all the 162 seats in East Pakistan (Bengal). With an overall majority under which he in theory could have ruled the whole of Pakistan, Mujib pushed hard for greater Bengali autonomy. The Bengalis, who had long felt themselves exploited, a pawn in Pakistan's power politics, were jubilant. According to the rules, Mujib should have become Prime Minister of all Pakistan, with Bhutto as leader of the opposition. Instead, with Yahya Khan increasingly sidelined, the highly visible Bhutto threatened any party member who went to Mujib's parliament in the East Pakistan capital of Dhaka, and presented himself as the saviour of West Pakistan. The army tried desperately to keep the two halves of the country together, but was underweight and overwhelmed in Bengal. To add to the mayhem, an Indian flight from Srinagar to Delhi was skyjacked, prompting India to deny Pakistan the right to overfly its territory. Flights from West to East Pakistan had to be diverted over Sri Lanka, a distance comparable to a transatlantic flight.

Against a background of flag burning and direct action against army units, innocent Bengali professionals and intellectuals were rounded up. Fearing a backlash, the army isolated its Bengali troops in West Pakistan, while reinforcements of West Pakistan troops were airlifted via the long route eastwards. As waves of Bengali refugees from East Pakistan arrived in India, the conflict escalated into outright war between the two giants of the subcontinent. But the Pakistani garrison troops in the east of their country were no match for Indian armour and air attacks from just across the border, while an Indian naval blockade of the Bay of Bengal added to the isolation. After direct Indian attacks on West Pakistan, the 93 000-strong Pakistan army in Bengal surrendered to Indian forces on 16 December 1971. The army, the proud backbone of Pakistan, was humiliated, and East Bengal declared itself the independent nation of Bangladesh.

In the ensuing vacuum in the rump of Pakistan, Bhutto declared himself president and martial law administrator, and began to stamp his ideal of Islamic socialism against a chaotic background of political assassinations, another uprising in Baluchistan, language status riots in Sindh and labour disturbances in Karachi. To provide a new focus and direct attention away from the chaos, Bhutto wanted to consolidate links between Pakistan and the populous and oil-rich nations of the Middle East. In February 1974, an Islamic summit at Lahore included Yasser Arafat representing Palestine, King Faisal of Saudi Arabia, Colonel Qaddafi of Libya, and Presidents Assad of Syria, Sadat of Egypt and Boumedienne of Algeria. To gain support in his own country, Bhutto introduced new measures to increase his country's Islamic profile and appease the powerful religious leaders. When the *Qaid-i-Azam*, Muhammad Ali Jinnah, had been elected founding President of Pakistan in 1947, he said 'You may belong to any religion or caste or creed, that has nothing to do with the business of the State'[3]. Jinnah had still seen the almost mystical power with which the nominally Islamic Mughal dynasty had been happily accepted by a Hindu majority. Bhutto discarded its final vestiges and took the step of making Pakistan officially Islamic. Alcohol was banned, the teaching of Arabic and the Holy Qur'an in schools emphasized, and more people were allowed to go on the *Haj* pilgrimage to Mecca. Because of the fundamental Islamic belief in the absolute and unqualified finality of Muhammad's prophethood, the Ahmadis, who had set their founder Mirza Ghulam Ahmad on a pedestal, became outcasts. On 7 September 1974, the Ahmadi community in Pakistan was officially declared 'non-Muslim'. Master politician Bhutto previously had no personal religious axe to grind against the Ahmadis, but probably saw them as a convenient pawn in his new game. Formal excommunication immediately sparked another round of anti-Ahmadi riots, and a pharmacy belonging to the Salam family in Multan was sacked.

Abdus Salam's diary entry for that day said 'declared non-Muslim, cannot cope'. It was a pit of despondency for Salam, who at the same time was trying to combat the US/Israeli embargo of his centre in Trieste. But nothing could shake his own faith. To underline his personal pride in being Muslim, he grew a beard and assumed the prophetic forename 'Muhammad'. At Trieste, he continued to lead holyday prayers in his room. For Salam's links with his home country, this was the last straw in a process that had begun several years earlier when the investment in science begun by Ayub Khan at Salam's suggestion largely evaporated in a move to rebuild Pakistan's military might after the 1971 debacle. With Bhutto pushing for the country's sophisticated peaceful nuclear programme to be refocused into a drive for weapons, under increasing pressure Salam's colleague and confidant Ishrat Usmani resigned as head of the nation's Atomic Energy Commission. He had already raised prominent eyebrows at an international conference on nuclear physics at Dhaka in 1967 when he referred to 'fossils in the Pakistan government'. Usmani was replaced by Munir Ahmad Khan, who Salam had met on his missions to IAEA headquarters in Vienna. Salam's role as scientific advisor had already been downgraded when the less scientifically aware Yahya Khan had taken over from Ayub Khan, but several days after Bhutto's excommunication of the Ahmadis, Salam tendered his resignation as an advisor: 'You are aware that I am a member of the Ahmadi community in Islam. I believe that the recent decision of the National Assembly in respect of this community is contradictory to the spirit of Islam because Islam does not give any segment of the Islamic community the right to pronounce on the faith of any other segment, faith being a matter between man and his creator.' The resignation was accepted by Bhutto, who nevertheless asked Salam to continue giving advice informally. 'This is all politics,' he tried to placate Salam, 'Give me time, I will change it.' Salam asked Bhutto to write down what he had just said on a note that would remain private. 'I can't do that,' replied the master politician. Despite his disappointment and disillusionment, the ever-ebullient Salam proposed setting up a series of summer schools 'Physics and Contemporary Needs' in Pakistan to bring distinguished scientists to the country, and with help from his International Centre in Trieste. These schools began in 1976 in Nathiagali, in the Murree Hills to the north of Islamabad, and continue under the aegis of the Pakistan Atomic Energy Commission. Another move was the Asian BC-SPIN (Bangladesh, China, Sri Lanka, Pakistan, India, Nepal) physics schools.

With the two men, Salam and Usmani, who had fostered Pakistan's nuclear expertise and its peaceful application now gone, the nation was firmly set on a new nuclear route. In the wake of the defeat by India and the creation of Bangladesh, Bhutto's vision was that Pakistan could realign its allegiances more towards the Muslim countries of West Asia and North Africa. With this new axis, his plan was also that Pakistan would also become the first Muslim country to have an atomic bomb. But he was not the only Muslim leader who had this dream. 2500 kilometres to the west, Saddam Hussein in Iraq had similar ideas. The Iraq Atomic Energy Commission had created a strong team of scientists who had been nuclear-educated in the West. In 1974 the Iraq AEC negotiated the purchase of a reactor from France, the better to produce bomb-grade plutonium. Whatever else it is up to, any Atomic Energy Commission has to have a politically correct main entrance, and in April 1975 the Iraq AEC organized its first scientific conference on 'Peaceful Uses of Atomic Energy for Scientific and Economic Development', held in Baghdad. Salam spoke on the 'Unification of Fundamental Forces and Elementary Particle Physics', a showcase for his latest ideas developed in collaboration with the Indian physicist Jogesh Pati. With few in Baghdad having any appreciation of the subject, this must have been another incomprehensible scientific talk. He also chaired a session entitled 'Growth, Energy Needs and Environmental Aspects'. At the meeting, Salam met key members of the Iraq nuclear effort. Despite the openness of the meeting, Salam knew what was going on behind the scenes[4]. He was concerned by Jaffar Jaffar's subsequent imprisonment in Iraq, which became the focus of Western attention and sympathy. Khidir Hamza was a visitor to ICTP Trieste, where he regularly met Salam and enjoyed a respite from the relentless pressure of Baghdad's bomb commitment.

Bhutto was now looking for knowhow to head Pakistan's new bomb effort. Such skills would not come from academia. His gaze fell on Abdul Qadir Khan, a nuclear engineer working in the Netherlands for the Dutch partner of the European uranium enrichment centrifuge consortium URENCO. Born in Bhopal in British India in 1936, A. Q. Khan, a Muslim, did not migrate to Pakistan until 1952, where he was educated in Karachi before studying at universities in Germany, the Netherlands and Belgium, obtaining a doctorate in metallurgy. Knowledgeable in the techniques of uranium enrichment, he was spotted by Bhutto and moved with his family to Pakistan in 1976. There he masterminded the nation's nascent bomb programme.

On 7 June 1981, Israeli jets screamed out of the desert and bombed the gleaming Osirak reactor at the Iraqi al-Tuwaitha centre. If this did not actually destroy Iraq's bomb plans, it severely hampered them. This left Bhutto and Pakistan alone on the nuclear road that led to the detonation of Pakistan's first nuclear bomb. But Ali Bhutto was not there to see it. The next great upheaval in Pakistan politics had come in 1977 in an army coup headed by General Mohammad Zia ul-Haq, and in 1979 Bhutto was hanged on a charge of complicity in murder. His daughter Benazir bravely took up the tattered standard of the Pakistan People's Party. Behind the politics, the nation's new nuclear effort under A. Q. Khan pushed ahead despite all the intervening regime changes, and on 28 May 1998 in Chagai, near the Afghan border, the Raskoh mountains trembled. Instead of the traditional countdown, the detonation button was pressed with the cry *Allahu akbar* – God is great.

While Bhutto had banned alcohol and excommunicated Ahmadis, Zia was a religious zealot who introduced Islamic *sharia* law and exploited Islamic fervour to affect what was happening in neighbouring Afghanistan. Among many other things he drastically reinforced the anti-Ahmadi measures introduced by Bhutto, decreeing that the movement could no longer use Islamic symbols or even describe itself as Islamic. New application forms for official documents required applicants to state their religion, where 'Ahmadi' was listed, along with 'Muslim', 'Christian' or 'Hindu', as a separate creed, implying that anyone who stipulates 'Ahmadi' is automatically not Muslim. The religion is then stamped in the passport. Applicants describing themselves as 'Muslim' are moreover required to sign a statement declaring that they are not Ahmadi, that Mirza Ghulam Ahmad is 'an impostor', and that his followers are non-Muslim. Every Pakistani Muslim adult has to make these declarations. Extremists called for even more severe measures. Ahmadi publications mailed to emigrants arrived scrawled with hate messages. Ahmadi congregations were not allowed to use a loudspeaker to broadcast *azhan*, the muezzin's strident call to prayer. Innocent Ahmadis were jailed for sedition, or even shot or knifed in their homes or in the street[5]. Like many other religious movements in history that had fallen victim to such intolerance in their homeland, the Ahmadi community looked out towards the world, but could never forget its Punjabi origins.

With his ties to Pakistan loosened, Salam turned his ambitions and political talents elsewhere. The United Nations was a natural target. After his introduction to the UN by Zafrullah Khan, his involvement in the Atoms for Peace programme in the mid-1950s and his subsequent involvement in the Pakistani delegation at the International Atomic Energy Authority, in 1964 he was appointed as one of the experts on the UN's Advisory Committee on the Application of Science and Technology, and in 1971–2 served as its Chairman. This committee provided input for the UN's Committee on Science and Technology for Development, an important part of the UN's goal of using science and technology as a catalyst to help less-developed countries.

UNESCO (the United Nations Educational, Scientific and Cultural Organization) was one of the first specialized agencies of the United Nations to be established. Its stated objective is to contribute to peace and security by promoting international collaboration through education, science, and culture in order to further universal respect for justice, the rule of law, and the human rights and fundamental freedoms proclaimed in the UN Charter. Initially, the infant agency was called the United Nations Education and Cultural Organization, but nuclear detonations brought a growing awareness of the importance of science and the 'S' was soon added to the acronym. Its first Director General was Julian Huxley, the distinguished British biologist and humanist, and grandson of Thomas Huxley, one of the foremost advocates of Darwin's theory of evolution. Julian Huxley had extended these principles to cover political and social contexts, and was a highly visible spokesman for UNESCO's newly added 'S'. Along with the IAEA, UNESCO was one of the founding pillars supporting Salam's Trieste Centre, and by the 1970s was contributing significantly to its budget. However, UNESCO's cultural objectives were not always easy to reconcile with its parallel commitments to human rights and fundamental freedom, and its business could be influenced by transient political questions. As explained in a previous chapter, in an atmosphere clouded by the 1973 war between Israel and its neighbours, a series of resolutions by the UNESCO General Assembly effectively excluded Israel from the organization, with the backlash having immediate consequences for the Trieste centre.

Before the internet, the World Wide Web and satellite TV changed the face of publication and broadcasting, communication was dominated by the printed word, where the world press and media still reflected a tradition of imperial and colonial power, particularly by Europe and its economic heirs in North America. To open up this restricted media world, UNESCO turned its attention in the late 1970s and early 1980s to what was called 'The New World Information and Communication Order', and a panel chaired by 1974 Nobel Peace Prize winner Sean MacBride was commissioned to make recommendations aimed at making global media coverage more objective, while promoting peace and understanding. It was an honourable but difficult task. The MacBride

Commission's 1980 report 'Many Voices, One World' identified a contemporary concentration and commercialization of the media on one hand, and the difficulty of access to information and communication on the other. The world of media had not yet emerged from its colonial past. The report called for democratization of communication and access to information, giving increased autonomy to national media: in other words, freedom of the press. While the report received broad and enthusiastic international support, ironically it was condemned by the United States and Britain as an attack on 'press freedom', where 'press' now inferred traditional powerful Anglo-Saxon outlets. Aggravated by other developments, with UNESCO increasingly being viewed by these countries as a Trojan horse for dissidents, they withdrew from UNESCO in protest. Whatever the sentiments of its other members, this was a major financial blow for the organization. With its budget severely cut, UNESCO had to trim its spending and undergo reforms before the US and Britain finally rejoined, the US in 2003 and Britain in 1997.

UNESCO's Director General during this difficult period was the Senegalese administrator Amadou-Mahtar M'Bow. Born in 1921, he completed his higher education in Paris before returning to Senegal, then under French rule, to teach history and geography. After directing the country's basic education from 1952 to 1957, he became Minister of Education and Culture as his country emerged as an autonomous republic within the 'French Community' period of internal autonomy (1957–1958), but resigned to help in the struggle for full independence. After this had been achieved in 1960, he became the country's first Minister of Education and then of Cultural and Youth Affairs. He was elected to UNESCO's Executive Board in 1966 and became Assistant Director General for Education in 1970. Appointed UNESCO's Director General in 1974[6], the first African to hold the post, he was visible and influential, and was reappointed for a second term in 1980.

As the end of M'Bow's second mandate neared, the question of his successor loomed. The veteran M'Bow was seen in many quarters as being too closely associated with a difficult period of UNESCO history. But the organization was in a predicament and needed a firm hand. A candidate from the Third World would be especially valuable, a counterweight to traditional geopolitical pressures, but on the other hand the candidate should still have the stature to reflect a globally equitable view. Abdus Salam was such a figure. It was a position that appeared to suit him in the twilight of his research career. He had a Nobel Prize.

He had established an international centre under UN sponsorship and had run it for more than a decade: he knew the politics of education (E), science (S) and culture (C), so was a living example of what the UNESCO acronym stood for: he bridged international barriers, and was admired by the market economies of the West, the socialists of the East, the rich countries of the North and the poor of the South alike; he had served on and even chaired an important UN Advisory Committee; in 1981 he had been elected chairman of UNESCO's Advisory Panel on Science, Technology and Society.

Carefully, Salam had prepared his candidature, soliciting support from governments in every continent, except Australia, which had its own candidate. Italy, his new adopted home, was a staunch supporter. To provide a platform, Salam had drafted plans for UNESCO's reform, proposing to return to the scientific theme that Julian Huxley had pioneered for the organization, pushing the importance of scientific education, and the science and technology grounding for economic development that Salam held so dear. There were other candidates for the job, including former Canadian Prime Minister Pierre Trudeau, but international fame is not the main criterion when it comes to electing UNESCO Directors General.

Tirelessly, Salam travelled to some 30 countries in 1987. Despite all this lobbying, there remained one major barrier. Salam had to have the support of his home country. Although, as an Ahmadi, he had been excommunicated from Pakistan for more than a decade, he had never given up his nationality. Salam knew that it was vital to have Pakistan's formal backing. He had supported other people's bids to obtain UNESCO posts and had learned the rules of the game. However, under the Islamic revivalist President Zia ul-Haq, Pakistan's formal candidate for the UNESCO Director Generalship was Lieutenant General Sahibzada Yaqub Khan, former Governor of East Pakistan, one-time Pakistan's Ambassador to Washington, and then Foreign Minister. Yaqub Khan's candidacy was lobbied by Attiya Enayatullah, who reminded the Paris assembly that just as a general had saved France, so another general would save UNESCO[7]. Salam had a counterplan. To stage manage Pakistan's emergence from another period of rule under martial law without challenging his own leadership, Zia ul-Haq had appointed Muhammad Khan Junejo as Prime Minister. Salam proposed to Junejo that Pakistan could have two candidates. A Director Generalship did not necessarily call for an army general, and Pakistan could also propose a scientist-administrator. Knowing that Zia's approval would nevertheless be required, Junejo demurred.

To elect the new Director General, in October 1987 the 50-member UNESCO Executive Council had to reach a simple majority decision, with candidates eliminated in successive rounds of voting. Yaqub Khan, who had been supported by several Asian nations, dropped out after two rounds. Although the radical incumbent M'Bow had been alienated by many Western delegations, he still had a substantial following in the Executive Council. Sensing danger if Yaqub Khan supporters switched to M'Bow, the floating vote lined up behind Federico Mayor Zaragoza of Spain, a former Deputy Director General. Seen as a compromise candidate by many Western nations, he had been gaining support as the voting progressed. He was duly elected as UNESCO's eighth Director General. It was a turning point in the history of UNESCO, but the episode was a bitter blow for Salam, an also-ran in the race, his ambition in tatters. The importance of science for developing nations was not to be pushed to the front of the political stage. But some who knew Salam quietly breathed a sigh of relief. They felt that running a small institute where Salam knew everyone by name was one thing, but piloting a huge and ungainly vessel through difficult international waters was a totally different affair. The administration of UNESCO could never have been compressed into a few hours each day.

Had the United Kingdom not temporarily quit UNESCO, it would have been another possible sponsor for Salam's bid to become the organization's Director General. Prime Minister Margaret Thatcher told Salam that if his UNESCO bid were successful, the UK would return to the organization. Even before the Nobel award, Salam had become a major voice in UK subnuclear physics. Following the successful establishment of the Trieste centre, he and Paul Matthews carefully constructed a plan for a British National Theoretical Physics Centre, to be built at Imperial College. A detailed architect's proposal was prepared, but the scheme foundered as the British scientific community at large struggled to promote, if not just to safeguard, their share of the research funding cake. 1981 marked the 150th anniversary of the birth of James Clerk Maxwell, Britain's greatest physicist after Newton, but who has received scant national recognition. Salam was one of the few in Britain who remembered Maxwell, but his efforts came too late to produce visible results. Ironically the anniversary was marked in the Soviet Union.

The visibility of the expenditure on subnuclear science in Britain that had torpedoed Salam's plan for a national theoretical physics centre at Imperial College was being thrown into sharper focus. In the 1970s, it had been decided to streamline research effort by discontinuing the home-based national programme in experimental elementary particle physics and to concentrate instead on UK participation in the pan-European work at CERN in Geneva. However, instead of streamlining research, this move highlighted the cost of contributing to CERN's budget. Although this was shared by CERN's West European Member States, the UK was traditionally one of the major contributors, siphoning off a fat sum that other scientists enviously saw as detracting from their own funding. In 1984, the British Advisory Board of the Research Councils commissioned a review group to look into look into UK participation in particle physics in general and international collaboration programmes in particular. The report, published in 1985, recommended that continued UK membership of CERN should be dependent on CERN cutting costs. Salam had given evidence to the review group, led by Sir John Kendrew (the molecular biologist who earned the 1962 Nobel Chemistry prize for his work on the structure of myoglobin), but was a vociferous critic of its recommendations. In a characteristically well-researched article – 'Particle Physics: Will Britain Kill its Own Creation,' – illustrated with quotes from P. G. Wodehouse, published in the *New Scientist* on 3 January 1985, Salam concluded 'Clearly, a country that has upheld fundamental science . . . cannot lightly absolve itself and withdraw from supporting this most exciting adventure of ideas of our times'. The UK remained a member of a CERN that, like UNESCO, had to undergo cost-cutting therapy.

In 1981, Salam urged UK Prime Minister Margaret Thatcher to award a life peerage to Paul Dirac, then 79, and living in retirement at the University of Miami, for his contributions to British and world science. Nothing happened, but it is doubtful that the taciturn Dirac would have wanted the honour anyway. When he had heard the news of his 1934 Nobel Prize, Dirac had initially wondered about turning it down, not wanting the attendant publicity. Ernest Rutherford quickly pointed out that refusing the Nobel Prize would generate more, and worse, publicity.

The British honours system, like the Nobel prize, is admired in some quarters, belittled in others. But it serves a purpose, enabling contributions in all spheres to be acknowledged and recognized without having to dispense large amounts of money. The pinnacle of Salam's recognition in the UK came in 1989, when he was made an Honorary Knight Commander of the Order of the British Empire (KBE). The Order had been established by King George V earlier in the twentieth century specifically to recognize non-combatants who had helped the country in the First World War. It was initially a junior partner in Britain's honour system, which traditionally had concentrated on military and diplomatic achievement, but gained in status over the years. Although nominally making him a Knight, the award did not entitle Salam to use the accolade 'Sir', which is reserved for British citizens. Ironically, in earlier years, British Indians, like Muhammad Zufrullah Khan, had been able to assume the 'Sir' title as at that time they were British subjects. Salam received his honour at Buckingham Palace on 26 June 1989. He was the first non-British scientist to receive the award, although a similar honorary award had gone in 1914 to the Italian wireless pioneer Guglielmo Marconi, who did all his important work in the UK.

Robust young men do not worry about their health. Before today's awareness of the benefits of physical fitness, the body was considered by most simply as a vehicle to transport the brain. Salam's hyperactive brain consumed a vast amount of energy, which was supplied by frequent snacks. His pre-breakfast of nuts, fruit, cheese and biscuits after getting up and saying his dawn prayers was followed by a full breakfast after the first tranche of his day's work already had been done. Before becoming tangled up in whatever the day brought, breakfast was a good time to entertain visitors – Zafrullah Khan in London, or visiting notables in Trieste. When travelling, Salam always kept an ample supply of sweets, biscuits and nuts in his bag to nibble throughout the day between meals. As he moved into middle age, his youthful waistline began to swell. A sudden awareness that physical fitness might be beneficial led him to purchase a rowing machine that was delivered to Campion Road, Putney, but that remained unused.

The additional responsibilities of running his centre, with his life divided between Trieste and London, and with his official duties in Pakistan, was increasing his workload as he reached the age of forty in 1966. In London, he lived comfortably with his extended family. In Trieste, he lived in a tiny villa at the Centre and worked up to 18 hours a day. He neglected his own comfort and became plagued by throat problems, eventually alleviated by a tonsillectomy in London. A few years later, in 1968, he came down with appendicitis while in Trieste, but two days after the emergency operation was already conducting physics research sessions from his hospital bed.

Salam privately gauged his life's passage and progress with that of the Holy Prophet Muhammad. Born in Mecca in 570, Muhammad's career as a merchant was successful but otherwise unremarkable until the month of Ramadan, 610, when in a cave above Mecca he experienced the first revelation of the word of God. At the relatively advanced age of forty, Muhammad realized his calling as the divine messenger. While Muhammad had been a merchant by trade, Salam was a theoretical physicist: both had been successful in their chosen professions. Both would have had comfortable lives had destiny not intervened. Salam's calling was the creation and establishment of the Trieste centre, which he achieved at about the same stage of his life that the Holy Prophet had encountered his true destiny. Muhammad then began his bid to establish the faith, leading to the *hijra*, the flight from Mecca to Medina in 622 at the age of 52, considered the starting point of the Islamic calendar. Salam's Nobel Prize came when he was 53. At the age when the Prophet Muhammad was warring with the infidels in Mecca, Salam was battling with funding problems at Trieste and his excommunication by Pakistan. As if to underline the parallels, he awarded himself the forename Muhammad.

With Islam finally ascendant, Muhammad's return to Mecca in 632 established the final form of the ritual commemorated in the pilgrimage of the *haj*, but soon after his arrival, he died, aged 63. Salam was very aware of the life history of the prophet: one of his favourite books was the *Shamail-i-Tirmizi*, which Zafrullah Khan had translated into English years ago, adding the dedication 'with deep gratitude to Abdus Salam, eminent physicist, with whom the idea of this book originated'. In the mid-1980s, as his lifespan approached that of the Holy Prophet, Salam realized that his accomplishments too must be nearing their productive end. His bid to head UNESCO had come to nothing, and his attempts to further the cause of Islamic science had failed to make headway. The powerful drive of aspiration and achievement that had propelled him through life suddenly dropped away and left him isolated, without support. The abrupt stillness was unfamiliar and perturbing. Those around him noticed that his voice lost energy and his movements slowed, and attributed it just to the effect of age.

But physical problems soon became apparent that were unconnected with his neglect of the rowing machine. Salam began to have difficulty walking and had to resort to a stick. Some of his close colleagues at Trieste saw it as a manifestation of his supreme disappointment following his UNESCO bid, a new hopelessness that sapped his vital strength. But it was not merely a psychological blow. The unique myriad of molecules that provided the genius of Salam had a hidden flaw. His right thumb stiffened and hindered his writing. Initially considered a trivial muscular strain, it refused to clear up. After several falls negotiating the steep stairs up to his villa at Trieste, it was clear that something was wrong. If people fall, it is normally because they stumble and pitch forwards: Salam fell backwards. He went to see specialists at Johns Hopkins Hospital in Baltimore, stopping first at the house of the Pakistani theoretical physicist Qaisar Shafi, who worked at the nearby Bartol Institute of the University of Delaware. While a meal was being prepared, Salam picked up the autobiography of the Swedish actress Liv Ullman, and became so engrossed in it that he later took it with him to the hospital. During the meal, Amtul Hafeeza scolded her husband for eating too many sweet *ghulabjamuns*. The specialists at Johns Hopkins saw only a debilitation similar to Parkinson's disease and could draw no firm conclusion. As he left the hospital, Salam whispered a Punjabi insult at the doctors.

In November 1989, Salam attended the formal inauguration of CERN's new particle accelerator, LEP, housed in a 27-kilometre tunnel under the Swiss–French border near Geneva, where French President François Mitterrand was guest of honour. By then, Salam's difficulty in walking meant that he was confined to a wheelchair. Soon after, in January 1990, he attended the International Economic Forum at Davos. His mobility continued to deteriorate and was eventually diagnosed by Professor Andrew Lees at London's Hospital for Neurology and Neurosurgery as progressive supranuclear palsy (PSP), a neurological affliction due to tiny changes deep inside the brainstem. It affects about one person in a hundred thousand, but which had then only recently been identified, as it is often indistinguishable from other neurodegenerative conditions such as Parkinson's and Alzheimer's diseases. There is no known cure. PSP is not in itself fatal, but its victims, dumbfounded and tottering, usually succumb to accumulated side-effects such as injury, malnutrition, or pneumonia. Another victim of the disease was British entertainer Dudley Moore (1935–2002). Salam's falls became more frequent, and the rooms in his Trieste villa were fitted with protective cushions. Stoically, he carried on with his work as best he could, but it was becoming increasingly difficult for him to function. His speech deteriorated and he was often incomprehensible, but those that knew him saw that his brain continued to function, imprisoned in a body that no longer responded to its control system. Stubbornly, Salam continued to travel, despite the humiliating indignities it brought. In 1992 he was vice-president of the jury for the first award of UNESCO's Félix Houphouët-Boigny Peace Prize, named after the first President of the Ivory Coast. The prize went to F. W. De Klerk and Nelson Mandela for their work towards South African regime change. In 1993, he visited Bangladesh, where he helped lay the foundation stone for the International Centre for Science, Technology and Environment for Densely Populated Regions in Dhaka, another international venture with Trieste parentage. His speeches were inaudible, and the patient audience applauded politely.

He had always dictated his best prose, but scientific papers had to be drafted by hand. Now unable to write, even his scientific work had to be dictated. A helper came from Pakistan to look after his special needs in the villa at Trieste. Physics research became more sporadic, and, at least from the outside, appeared to dry up in 1993. Salam's friends, colleagues and admirers all over the world decided that a meeting in his honour should be organized while he was still capable of enjoying it. In March 1993, a conference 'Highlights of Particle and Condensed Matter Physics' at Trieste coincided with Salam's retirement as Professor of Theoretical Physics at London's Imperial College. Among the speakers at Trieste were long-time colleagues such as Steven Weinberg, who shared the Nobel Prize with Salam in 1979 and who had been his guest at Imperial College in the early 1960s, C. N. Yang, whose association with Salam went back even further, former collaborators Robert Delbourgo, Jogesh Pati, Seifallah Randjbar-Daemi, John Strathdee and Tom Kibble, former students such as Yuval Ne'eman, Riazuddin and Fayyazuddin, and admirers such as Edward Witten and Gerard 't Hooft. As the tributes flowed, Salam sat in his wheelchair at the back of the auditorium. The culmination of the event came when he was awarded an honorary doctorate by the University of St. Petersburg. The Rector had made the journey specially.

At such a 'Festschrift'[8] meeting, the contributions are highly subjective – some speakers use the opportunity to eulogize the guest of honour, others simply report their latest work, adding a salutatory tribute at some stage in their talk. Had Salam been able to add a vote of thanks, he would surely have been autobiographical. He had a keen sense of history, whether political, religious or scientific, which became more incisive as he grew older. At the inauguration of the definitive home of the Trieste centre in 1968, the programme of talks had included a series of biographical sketches by or of distinguished figures – 'From a Life of Physics'[9]. More than two decades later, in 1989, another major seminar at Trieste marked the 25th anniversary of the centre. This time it was the turn of Salam himself to speak himself on 'From a life of physics'[10]. This delineated his own career, building on what he had sketched in his Nobel address in 1979 and his 'Physics and the excellences of the life it brings' talk at a 1985 physics history symposium in Fermilab, near Chicago[11]. But this time Salam ironically reflected how his own abilities had blinded and baffled those whose advice he trusted, thereby misdirecting his early career. Had he realized, Salam said he would have aimed directly for a research career at Cambridge, rather than repeating his undergraduate mathematics. This had been pointed out to him in 1986 by Singapore's Prime Minister Lee Kuan Yew, who had been Salam's contemporary at Cambridge. It had taken a long time for the penny to drop.

With Salam effectively out of action, Trieste clearly needed someone else at its head, and in 1995 the Argentinian physicist Miguel Angelo Virasoro took over as the Centre's Director, after Salam had formally become its President in 1994. Although a very private person, Salam had always craved the company of his intellectual peers and enjoyed the tumult of politics. Now he had involuntarily retreated into silence and isolation. Salam retired to the UK, where most of the time he sat mutely in Louise Johnson's riverside house in Oxford, commuting with difficulty from the downstairs living room to his bedroom upstairs. A nurse came in during the day to look after him. Periodically he transferred to his home in Putney, London, where his son Ahmad and daughter-in-law Sophia lived. Recordings – whether readings from the Holy Qu'ran or operatic classics with the supreme power of Pavarotti's voice – were a great comfort to him.

As the disease advanced and stripped him of outward expression, Salam's intact core still absorbed all that went on around him. He diligently followed world affairs on BBC radio. His mind probably ranged much wider: nobody could tell. Ironically, the science from which he had drawn so much energy and that had served his vivid imagination had grown stale. There were few new discoveries to point the way forward. In a stage now bereft of fashion, perhaps he quietly returned to the deep mysteries of the inner structure of matter that had been the business of his lifetime. But for such research, he had almost always needed a close collaborator, a filter and resonator for his imagination. Now there could be no more collaborators. Eye contact and lengthy questioning were the only way to communicate: he could whisper single words and reply yes or no to questions, even to the end, but could not initiate a sentence. After visiting him, Luciano Bertocchi from Trieste said 'his body was no longer obeying his spirit', but that Salam's eyes sparkled when he heard news from Trieste.[12] Those soft eyes were the deep mirror of the feeling that still burned brightly within, despite all the obstacles and humiliations. These increased after the summer of 1995, when he had to be fed via a catheter.

A small event in Islamabad marked his 70th birthday on 29 January 1996. The opponents of the Ahmadi movement had tried to stop it: any function held to honour Salam would amount to defaming Pakistan, they trumpeted, but Prime Minister Benazir Bhutto, who on other occasions had snubbed Salam, rejected the call. Her congratulatory letter was delivered by Pakistan's High Commissioner in London to Salam in Oxford. Sensing the end, colleagues made the long journey to visit him. Munir Ahmad Khan saw him in August. Speaking in Punjabi, he conveyed belated 70th birthday greetings from Pakistan. '[Salam] listened but could not respond. He just stared at me as if he had risen above praise. As I departed, he pressed my hand feebly.'[13]. Seifallah Randjbar-Daemi visited him in October. Louise Johnson had been reading her husband some Persian poetry in translation. Randjbar-Daemi reported how his latest research with John Strathdee was progressing. At Trieste, Salam's collaborators were still collaborating: his eyes shone. Strathdee and Randjbar-Daemi had been trying to render the forces between subnuclear particles into a form that made computer simulation feasible. 'Don't forget gravity', Salam croaked. The sudden joy of scientific insight had broken the binary yes/no straitjacket that appeared to have restricted his utterances. Like Einstein, he was afraid to confront death without uncovering the ultimate theory that encompassed all of Nature's forces.

Towards the end, Salam's passivity was broken by perturbing fits. Several times he had injured himself in falls, sometimes calling for hospitalization, but in November 1996 he fell in his bedroom in Oxford one last time and never recovered consciousness. He died four days later, silently, on 21 November. Muslims believe that life is a transient, merely a preparation for the afterlife. Salam had had a long time to contemplate his passing. Transcendental concepts, such as consciousness and the soul, he believed to exist outside the realm of physics and elementary particles that had busied him during his lifetime. For him, paradise was very real – a spiritual state where one's soul would be liberated from debilitating emotions such as worry and fear. Once there, he yearned to re-establish contact with his father, who had firmly guided his early progress and taught him the Islamic faith[14].

Accompanied by nineteen members of family, his body was flown to Pakistan, arriving in the morning of 25 November, and taken to an Ahmadi mosque in Lahore, where admirers offered prayers and paid their last homage. In keeping with the official ostracism of Ahmadis, there had been no official government delegation to meet the body. Despite his lifetime's achievement and having guided Presidential thinking for a decade, the highest-ranking Pakistan official present was a superintendent of police. Later that day, the body was taken to the main Ahmadi community in Rabwah. People lined the road as the convoy arrived. Overnight, thousands of people from all over the country filed past to pay their last homage to an international figure, the greatest Pakistani scientist the world had known. After the last prayers were said at 10 am the next day, his body was laid to rest alongside the graves of his parents in a cemetery reserved for those who had contributed a tenth of their annual income to charity and the community, and had been judged by their Ahmadi peers to be pious Muslims. His tombstone was proudly inscribed 'Abdus Salam, the First Muslim Nobel Laureate'[15]. Soon the grave was visited by contemptuous outsiders and the inscription edited by an imperious hammer and chisel to read 'Abdus Salam, the First... Nobel Laureate', and daubed with black paint. In death, as in life, Abdus Salam was vilified in the country to which he had tried to contribute so much.

REFERENCES

1. Palit P.K. and Namboodiri P.K.S., *Pakistan's atomic bomb*, New Delhi, 1979
2. Ghani, A., *Abdus Salam: A Nobel Laureate from a Muslim country*, (Karachi, published privately, printed Ma'aref, 1982)
3. Quoted in Bhutto, B., *Daughter of the East* (London, Hamish Hamilton, 1988)

4. Khidir Hamza, *Saddam's bombmaker* (New York, Simon and Schuster, 2001)
5. Gualtieri, A., *The Ahmadis* (Montreal, McGill-Queen's University Press, 2004)
6. This UNESCO meeting also adopted the anti-Israel motions that led to the boycott of the Trieste cetre by Israeli and US scientists.
7. Khalid Hassan, Pakistan Daily Times, 26 November 2006
8. Ali, A., Ellis, J., Randbar-Daemi, S. (ed.) *Salamfestschrift* (Singapore, World Scientific, 1994)
9. Salam, A., Fonda, L., (ed.), *Contemporary Physics: Trieste Symposium 1968* (2 vols), (Vienna, IAEA, 1969)
10. Salam, A., *A life of physics*, in Cerderia, H. A., Lundqvist, S. O., (ed.) *Frontiers of physics, high technology and mathematics* (Singapore, World Scientific, 1990)
11. Salam, A., *Physics and the excellences of the life it brings*, in *Pions to quarks*, Brown, L., Dresden, M., Hoddeson, L., (ed.) (Cambridge, CUP, 1989)
12. Bertocchi, L., *My association with Abdus Salam*, in Hamende, A. M. (ed.), *Tribute to Abdus Salam*, (Trieste, ICTP, 1999)
13. Khan, M. A., *Lifelong friendship*, in Hamende, A. M. (ed.), *Tribute to Abdus Salam*, (Trieste, ICTP, 1999)
14. Vauthier, J. *Abdus Salam, un physicien* (Paris, Beauchesne, 1990), 96
15. The first Muslim to earn a Nobel prize was Egyptian President Anwar Sadat, who shared the Peace Prize with Israeli Prime Minister Menachim Begin in 1978, one year before Salam's award.

15

Prejudice and pride

Every October, the announcements of the year's Nobel prize awards are trumpeted in the Western world on prime-time TV and splashed across newspapers. They have become symbols of sovereignty, highlighting the emergence of new countries in cultural arenas, or underlining the prowess of established nations. Much prestige is at stake. Particularly for the science prizes, news editors have to struggle to decode the often obscure and complex developments that merit these accolades into something approaching comprehensibility. Specialists who have worked all their lives in obscurity become celebrities overnight, and are invited to participate in TV chat shows, and write columns for newspapers. In 1984, Carlo Rubbia shared the Nobel Physics award with Simon van der Meer of CERN for their work that culminated in the discovery of the W and Z carrier particles predicted by the electroweak theory. After learning the news, later that day Rubbia was in a taxi *en route* for Salam's Trieste institute. The driver had his radio tuned to the news broadcast. 'Hey,' he shouted, 'An Italian has won the Nobel prize!' 'Yes, it's me,' replied Rubbia laconically. 'Get moving.'[1] When Rubbia arrived, Salam, who ten years before had predicted what Rubbia had discovered, was there to greet him. So was a phalanx of Italian TV crews.

Twenty years later, the Egyptian scientist Ahmed Zewail was honoured with the Chemistry Prize for his studies using laser light to track chemical reactions, 'freeze-framing' their evolution by successive snapshots taken at femtosecond (10^{-15} s – a quadrillionth of a second) intervals. Immediately after his first degree at Alexandria University, and before moving to the US, Zewail worked as a demonstrator (*moeid*), teaching undergraduates and working towards his master's degree. He says 'As a *moeid,* I was unaware of the Nobel in the way I now see its impact in the West. We used to gather round the TV or read in the newspapers about the recognition of famous Egyptian scientists and writers by the President, and these moments gave me and my friends a great thrill.'[2] Like Salam, Zewail had to emigrate to the West to do his Nobel science.

The perception of Nobel Prizes, like that of beauty, is in the eye of the beholder. In the West, the awards are a pinnacle of achievement, and this stature makes them easily visible. But a mountain can appear very different from opposite sides. In 2003, Iranian lawyer Shirin Ebadi was awarded the Nobel Peace Prize 'for her efforts for democracy and human rights. She has focused especially on the struggle for the rights of women and children.' While people around the world applauded this recognition, others maintained that it was an insult to and part of a continuing conspiracy against Islam. In a statement carried by the Iranian *Jomhuri Eslami* newspaper, a group from a major seminary said 'The decision by the Western oppressive societies to award the prize to Ebadi was done in order to ridicule Islam.'[3] How can what is supposed to be one of the world's highest honours also be perceived as insult and ridicule?

The Nobel prizes were established as a showcase for merit. In his will, Alfred Bernhard Nobel stipulated that they should go to those who have 'conferred the greatest benefit to mankind'. Awarded for Literature, Physics, Chemistry, Medicine, Economics and Peace, the annual list reflects topical themes, but with more than a hundred years of tradition, the roll of prizewinners and their work also displays evolution and progress. Nobel's will also said 'in awarding the prizes no consideration whatever shall be given to the nationality of the candidates, but that the most worthy shall receive the prize'.

Shirin Ebadi is one of the few Muslims to have been so honoured. The first was Egyptian President Anwar Sadat, who shared the Peace Prize with Israeli Prime Minister Menachim Begin in 1978 for their unexpected Middle East peace overture. In 1981 Sadat was assassinated by Egyptian hard-liners who condemned his rapprochement with Israel. One year after Sadat's award, in 1979 Abdus Salam became the first Muslim to win a Nobel Science Prize, and the first Pakistani to win any Nobel. The achievement was greeted in the West with the customary apotheosis – a crop of fresh honours began to arrive from all over the world, and funding for his Trieste Institute was immediately boosted. But the accolade in his home country has been very different.

Subsequently, the 1988 Literature Prize went to the Egyptian writer Naguib Mahfouz (1911–2006), whose initial literary success in the 1960s and 1970s created a new hub of Arabic culture. This became overshadowed by his controversial *Awlad Haratina* (*Children of the Alley*) that was banned in much of the Arab world after reactionary Islamic scholars declared its portrayal of religious figures to be blasphemous. In the darkness of such bigotry, writers who can still write are deemed more dangerous than what they actually publish. In 1994 Mahfouz almost died after being knifed in the neck, and was left unable to work. In 1994 Palestinian leader Yasser Arafat shared the Peace Prize with Israel's Prime Minister Yitzhak Rabin and Foreign Minister Shimon Peres for their resolute but eventually futile efforts towards resolving the perennial Israel–Palestine conflict. Such a pairing of names that not that long before had been sworn enemies soon created a new conflict of its own, and in 1995 Rabin was assassinated in his own country, a macabre reflection of the Sadat episode.

On a less controversial note, in 2005, Mohamad ElBaradei, the Egyptian Director General of the International Atomic Energy Authority (IAEA), and the IAEA itself received the Peace Prize for their efforts in preventing nuclear energy from being used for military purposes and for promoting its safe use for peaceful aims. In 2006, Muhammad Yunus from Bangladesh received the Peace Prize for his idea of 'microcredits' – miniloans to help disadvantaged people haul themselves out of poverty.

The world's 800 million Muslims make up about ten per cent of the world's population, but have garnered just a handful of Nobel awards, many of them generating more controversy than honour. Jews make up a small fraction of one per cent of the world's population, but have won hundreds of Nobel prizes. This track record alone is enough to convince some Muslims that the Nobel dice are loaded. But why such disparity and dissent?

The West has grown to view the Orient from afar through a thick prism that distorts the transmitted image. The dominion of Islam has expanded and contracted in history, but its cradle in the Near East has endured for well over a thousand years. Throughout this time, the membrane between Islam and the West, inflamed by lack of understanding, has been rubbed raw by mutual hypersensitivity, and the ulcerated wound has periodically erupted. In his perceptive book *The Last Mughal*, William Dalrymple writes 'The histories of Islamic fundamentalism and European imperialism have very often been closely, and dangerously, intertwined. In a curious but very concrete way, the fundamentalists of both faiths have needed each other to reinforce each other's prejudices and hatreds. The venom of one provides the lifeblood of the other.'[4]

Christianity, Judaism and Islam were nurtured in close proximity, share common scriptures, and variously acknowledge the key roles played by patriarchal figures. In the first millennium, feuds marked the emergence of Christianity and then Islam, but the vast counterweight of Christian Byzantium remained a stabilizing factor, until warlike Turkish invaders from further east upset a fragile equilibrium. Ostensibly to safeguard the interests of Western European pilgrims, Crusaders marched in with little understanding and even less concern for the local population. Hot-headed Frankish ignoramuses found it difficult to tell the difference between a Turk and an Arab, or to distinguish between a Muslim, a Jew, and an oriental Christian, and struck out blindly at all. A climax was the capture by the Crusaders of Jerusalem in 1099. In his *History of the crusades*[5], Stephen Runciman says 'The massacre at Jerusalem profoundly impressed the world. No one can say how many victims it involved; but it emptied Jerusalem of its Muslim and Jewish inhabitants. Many even of the Christians were horrified by what had been done; and among the Muslims, who had been ready hitherto to accept the [Crusaders] as another factor in the tangled politics of the time, there was henceforth a clear determination that [the Crusaders] must be driven out. It was this bloodthirsty proof of Christian fanaticism that recreated the fanaticism of Islam.'

The Crusader period also coincided with the apogee of Islamic culture. While the armies of the Cross, driven from the Levant, spent their frustration on fellow Christians in Constantinople or trampled into North Africa, Salam's avatar Michael the Scot was translating the wisdom of Arabic books into Latin.

A thousand years later, Western universities have distinguished faculties, even entire institutions, dedicated to the understanding of the East. Such expertise is in wide demand in politics, administration, commerce and the media. But from a reciprocal viewpoint these studies can be seen, like the Nobel awards, as patronizing and biased. The picture comes into sharp focus in Edward Said's influential book *Orientalism*[6], which contends that oriental studies in the West have been geared to an imperial colonialist tradition, the entire Orient being seen merely as an instrument to be manipulated. In his monumental work, Said claims the West observes the Orient from afar, and to some extent from above, through a thick window of selfish interest, and openly scorns the work of many 'orientalists' whose authority is elsewhere much esteemed.

One root of this subjective Western tradition of Orientalism, Said contended, was Napoleon's expedition to Egypt in 1798. Galled by resistance to his bid to conquer Europe, and piqued by British colonial successes in India, Napoleon invaded the Nile delta in 1798. An immediate goal was to drive a wedge between Britain and its Indian empire, and his ultimate dream was to emulate another European, Alexander the Great, who had wrenched Egypt from the Pharaohs and marched onwards to the Punjab. However, after the British fleet under Nelson destroyed Napoleon's supply lines at the Battle of the Nile, the isolated French army eventually had to retreat, as the Crusaders had done five hundred years before. Despite this failure, this expedition was nevertheless a turning point in the history of the Middle East. It was the first time since the Crusades that Western European armies had set foot in The Levant. Napoleon's uninvited appearance in the Orient set the scene for continued western meddling over the next two hundred years, and that continues to this day to be a source of conflict.

Accompanying the Napoleonic army was a specially commissioned team of scientists assigned to interpret the country and its culture, an unusual and imaginative move by a military leader. Its findings were published in the massive *Description de l'Egypte*, which appeared as 23 separate volumes from 1808 to 1823 under the general direction of mathematician Jean-Baptiste-Joseph Fourier (perhaps better known for his analytical theory of heat, and the 'Fourier expansion', a standard tool of modern mathematical analysis). The expedition had strong parallels with that of al-Biruni nine hundred years earlier, who had been commissioned to carry out a scientific description of India for the Muslim invader Mahmud the Ghaznivid.

Napoleon's failure meant that Britain could continue its bid to rule India. Large and complex, with immense resources of all kinds, India developed into a distinct British imperial unit, with its own administration, civil service and army. Odd fragments of Indian culture were even absorbed into British lifestyle. By the end of the twentieth century, one of the most widely appreciated dishes in Britain was *tandoori* chicken, a recipe invented after the 1947 partition of India, when Hindu refugees from the Punjab fled to Delhi and, with no other source of income, set up snack stalls in the street. Indian food writer Madhur Jaffrey relates how the spiced, succulent flesh took the city by storm.

Islam had first prospered in the Indian subcontinent under the Delhi sultanate, and later the great Mughal empire, which flowered for more than two hundred years and became a template for Indian culture. But the British steadily consolidated their hold until 1857, when the shock and affront of the Revolution/Mutiny was to them in the nineteenth century what 11 September 2001 became to the United States of America in a new millennium. From their lofty perch, the underlying reasons for Indian national fervour and the accumulated hatred for its colonial masters were largely invisible to the British. Faced with insurrection, the customary Western medicine was, and still is, a stern dose of repression – 'to teach them a lesson'. Muslims, already disoriented by the erosion of their traditional status by the colonial administration, were widely blamed for the uprising. The vestige of the once-proud Mughal dynasty, which nevertheless still stood as a common figurehead for the diverse cultures of the subcontinent, became a scapegoat. The furious initial British backlash, documented for all time by the poet Ghalib, was bloody and terrible. With the Emperor humiliated, India began to fall apart, and Islam in the subcontinent was on its knees.

Thus was the scene set for a Muslim–Hindu schism and the eventual fission of British India into two divergent nations. Unsure which direction to take after the sudden death of its guide and founding father, Muhammad Ali Jinnah, the new Muslim state of Pakistan did not show the tolerance which the Mughal emperors had shown centuries earlier. For Muslims, idol-worship – *shirk* – is a major sin, but in British India almost every day they had had to turn the other way as their Hindu compatriots paraded likenesses of Shiva and Vishnu through the streets. In Islamic Pakistan, frustrations that had been repressed for centuries were released. With Britons, Hindus and Sikhs all departed, the pacifist Ahmadi minority floated to the surface to become a convenient target for the vitriol that had first overflowed in 1857. Meanwhile India, predominantly Hindu but nominally secular, is officially more tolerant of the religions in its midst. There have been terrible mob riots and killings, but Sikhs and Muslims have been appointed to high office. The irony is that any Ahmadis who remained in India escaped the difficulties that their co-believers met in Pakistan.

In such sensitive political and religious chemistry, the highly visible Nobel Prize is vulnerable. With Pakistan under the orthodox Islamic regime of President Zia-ul-Haq, the status of Ahmadis in that country had reached rock bottom. However, Zia-ul-Haq, as a statesman sitting uncomfortably on a fence of his own making, recognized the international importance of Salam's award, and invited him to Pakistan as a state guest and to receive the nation's prestigious *Nishan-e-Imtiaz* (National Distinction) award. Whatever else Zia-ul-Haq had done, Salam, who had hated Bhutto, was happy at the chance to deal with the strongman who had deposed him and might now unlock doors. Islamabad's Qaid-i-Azam University decided to mark the visit by awarding Salam an honorary doctorate. However, the organizers did not dare to hold the event at the university, where it would certainly attract anti-Ahmadi right-wing students, whose religious fervour had widespread tacit support. Instead, the ceremony was held inside the security of the National Assembly, where access could be tightly controlled. Fayyazuddin, acting as Dean, wrote the citation, and Zia-ul-Haq himself conferred the degree. It was a difficult moment for the President/General, who knew that he could not afford to upset the powerful factions that had been jubilant at his continued anti-Ahmadi stand. To appease the hardliners, careful editing of the TV coverage improved Zia's image to the detriment of Salam[7]. In his talk, Salam said that he was proud to be the first Muslim scientist to be awarded the Nobel Prize. Far from echoing the claim, revivalist Muslim voices criticized the declaration as a desperate attempt to restore Ahmadi credibility. In a grotesque eructation of prejudice and hate, at the *Eid* service at the Lal Masjid mosque in Islamabad, Salam's Nobel award was scorned as a warped panegyric from the enemies of Islam.

Far from gaining widespread acclaim in his home country, Salam's 1979 Nobel award and even the Nobel Foundation itself became the target for vituperative criticism and outright condemnation. Pakistani revivalist schools saw the Nobel tradition as a 'Qadiani–Jewish lobby'[8] that deliberately ignores Muslim contributions, and that the Prize was deliberately awarded to Salam in 1979, the centenary of Einstein's birth, because 'Qadianis have a proper mission operating in Israel'[9]. Moreover, the Prize is basically sinful because the prize money is generated through the accrued interest on Nobel's capital, a financial practice unlawful in Islam – 'Those that revert to the practice of usury are the fellows of the fire, where they shall live for long'[10]. That Pakistan only developed the atomic bomb after Salam's departure as Presidential scientific advisor clearly demonstrated his 'incompetence' and his 'enmity'[11] towards his country. Later, Salam was frequently cold-shouldered in Pakistan. In 1988, after patiently waiting for two days in Islamabad to meet Prime Minister Benazir Bhutto, he was snubbed.

Salam's national humiliation is further illuminated by comparing his case with that of another mathematical prodigy of the early twentieth century Indian subcontinent – Srinivasa Ramanujan. Both men altered the course of modern science, and are well known throughout the world, but their respective legacies in their own countries are very different. Away from Government College University, Lahore, where there is a Chair of Physics and a hall named after him, it is difficult to find any acknowledgement of Salam in Pakistan. However, in 2007, a new Abdus Salam endowed chair in theoretical physics was created at Lahore University of Management Sciences. In another exception to his obfuscation, on 21 November 1998 the presses thundered out half a million two-rupee postage stamps bearing Salam's portrait, under the heading 'Scientists of Pakistan'. Honorary doctorates and elections to national academies and learned societies are a traditional mark of recognition. In India, Salam went on to collect five honorary degrees; in the UK six, and eight from African universities[12]. In Pakistan, Salam garnered just two: one, in 1957, came from the University of Punjab, when he was in the ascendant; the second was the carefully orchestrated event in 1979 after the Nobel Prize.

In 1987 the centennial of the birth of Ramanujan was marked in Kumbakonam, the Indian mathematician's hometown, by a procession through the streets. An elephant bore a life-sized, flower-garlanded portrait of the genius, wearing his Cambridge mortarboard cap. In Madras, Prime Minister Rajiv Gandhi signed the first copy of the latest edition of Ramanujan writings and presented it to Ramanujan's widow, Janaki.

When Ramanujan died of tuberculosis in 1920, age 32, his fame was mainly restricted to Tamil-speaking South India. Elsewhere in the world his name was known only to those who read mathematics journals. More than half a century had to pass before the value of some of Ramanujan's conjectures was realized. Now, people across the world know the name, just as they know the name of Albert Einstein, even if they do not fully, or even remotely, understand what these men did. The romantic story of a boy from a backwater of South India whose meteoric intellect went on to warrant world recognition is a continual source of inspiration for books and TV films. Ramanujan prizes and awards encourage and reward mathematical achievement (one is administered by the Abdus Salam International Centre in Trieste). In his homeland, the legend went on to inspire new generations of talent, but also became an embarrassment to the country that had initially neglected him. Ramanujan's genius had languished until his own stubborn efforts eventually brought him to the attention of an influential mathematician in faraway Cambridge. Ramanujan had known that there was no point in trying to sell his genius in his own country. After the tragedy of his short life, India assumed a collective guilt at not providing enough opportunities to nurture her own talent. Today it is different: wannabe Ramanujans are no longer neglected and left to fend for themselves. Indian research and development institutes are recognized throughout the world, and the country has become one of the world's leading centres of information technology.

Salam's arid Punjab was a very different part of British India from the jungle-green Tamil Nadu of Ramanujan. Punjab was predominantly Muslim, and had its own language: Kumbakonam was steeped in Hindu Brahmin traditions, and spoke Tamil. The family backgrounds were also very dissimilar. Ramanujan's father was a shop clerk, a shadowy figure who returned home tired after working long hours. His mother was his main influence, but herself was no intellect. Salam's father, Muhammad Hussain, was a schoolteacher, already a lofty status in a country where most of the population was illiterate. Both Salam and Ramanujan scored impressive marks in their school examinations and marched steadily up the ladder to their respective local Government Colleges, Salam in Jhang, Ramanujan in Kumbakonam, with modest scholarships. Their speciality, mathematics, unlike music or writing, is a talent that is difficult to communicate, and impresses by its contrariety. It is seen by most as intellectual medicine, to be swallowed rather than savoured. Ramanujan withdrew, sank ever deeper into mathematics, and became an oddity and a dropout. Only Godfrey Hardy heard his cries and rescued him from a swamp of oblivion.

In contrast, Salam was a polymath and a man of action. Initially, he compensated for the impenetrability of his mathematical talent by writing articles in English for his college magazine and by penning Urdu poems. His career did not languish, at least not initially, as he progressed from Jhang, to Lahore, to Cambridge, to Princeton. Only when he returned to Lahore in 1951 did his upward momentum begin to stall. His abilities had reached as far as the new country of Pakistan could provide, but he still aspired to go higher. With no oxygen to support him in his intellectual stratosphere, Salam left his homeland, a high price to pay, both for himself and ultimately for his country. But he did not want later generations to have to pay so dearly, and yearned for the new nation, and others like it, to aspire to scientific excellence. For this, he conceived an ambitious plan, wrote its scenario, and proceeded to act it out.

A long countdown started, but initially there was not enough thrust for it to rise from the ground until Paolo Budinich saw what Salam's plan had to offer Trieste. With this boost, Mission Salam could finally blast off and go into orbit. Budinich was for Salam the chance godparent that Hardy had been to Ramanujan. Without the help of Budinich and Trieste, Salam's scheme might have happened somewhere else – Copenhagen, Ankara, Lahore – but would have been different. It might not have happened at all, and the Third World would not have its oasis of research excellence.

(Curiously, Salam's home in Campion Road, Putney, was across the way from Colinette Road, where the sick Ramanujan had spent his last weeks in England in 1919 before returning to India.)

How did Abdus Salam achieve these objectives? He had an uncanny ability to understand a political environment, to see what needed to be done, to ascertain which people he did not know needed to be convinced, or those he did know persuaded. He did this by suggestion: he did not bark orders. As well as a shrewd politician, he was also a man of passion: whatever caught his imagination was quickly transformed into forward momentum. As his passion burned, it released energy. He would usually go to bed before 9 pm, but would rise at about 3.30 am and say his first prayers of the day before commencing work. These pre-dawn labours, often his most productive period of the day, would be fuelled by food and drink carefully set out the previous evening. By dawn he had already accomplished a lot. Robert Delbourgo recalls an impatient Salam phoning at six in the morning to enquire how work was progressing. In London, this was done in his personal room in Putney, always several degrees warmer than the rest of the house, heavy with the scent of incense sticks and with recordings of Holy Qur'an recitations in the background.

As his aspirations bore fruit, Salam became inundated by peripheral invitations, to address meetings and attend formal banquets. At these events, so many people would want to talk to him that he had no time to eat the elaborate dishes served, and took the precaution of eating at home first. For their book *The second creation*, Robert Crease and Charles Mann witnessed a banquet given in Salam's honour in New York City: 'Salam was given no time to eat; besieged by books to sign, hands to shake, babies to kiss, and youths eager for career advice, he spent the meal administering to each entreaty with unflappable reserve'[13].

Salam dealt with his workload through remorseless use of the time available, sandwiching research sessions between meetings and vice versa. He knew the value of contact with teachers and enlightened study, and recommended students 'not to waste time in playing cards or going on strikes or watching useless movies'[14]. Total concentration by the intellectually gifted can be physically debilitating, like running a race. After remaining in his study for days at a time and forgetting to eat meals, Newton was thin and grey-haired in his thirties. When Freeman Dyson carved out his milestone contribution on quantum electrodynamics, he wrote to his parents that the concentration 'nearly killed him'[15]. While he was still a student, Richard Feynman's dormitory neighbours found him rolling on the floor as his body was consumed by cerebral energy[16]. Salam's total concentration could make him forget what he was eating, a half-chewed morsel falling from his mouth. But more often it just gave the impression of absent-mindedness, as though parts of the brain normally available for everyday matters had been commandeered for higher things. Once the family went out and left him to babysit his infant daughter Sayyeeda, and returned to find the baby asleep, but covered in newspaper cuttings. Salam, oblivious to his daughter, had been assembling material from back issues of the *Economist* to illustrate a talk.

But at other times, his accessibility was legendary. Many theorists adopt the habit of leaving their office door open while they are working. This gives the corridors of prestigious university departments the aspect of a menagerie, where impressionable students can wonder at intellectual 'exhibits', in the same way that young post-war researchers had watched Einstein walking on the grass at Princeton. Leaving the door open invited casual callers, and students were impressed at Salam's ability to drop whatever he was doing and answer their questions, with no sign of irritation. The brief session concluded, he would immediately return to his work as though nothing had happened.

Particularly with the establishment and administration of his Trieste institute, travel took up a lot of Salam's time. He was continually commuting between his two jobs at Trieste and Imperial College,

London. Zufrullah Khan had advised the young Salam that travel should be used to broaden the mind, but this soon reached saturation. Salam was a humble and patient traveller, using economy class before the advent of business class, and frequently carrying excess baggage: documents; transparencies to illustrate his talks; books bought on his travels; home-cooked frozen food for his Trieste villa. In 1979, his Nobel Prize year, Salam's travel itinerary seemed to pass some watershed. He was increasingly invited to give lectures, sit on advisory committees, receive honorary awards, ... all of which involved extra air miles. With planning and concentration, work can be compressed into the time available for it, but travel has to run at its own pace. With air travel, once a privilege of money and power, becoming more accessible, planes and airport lounges were full and uncomfortable.

To rationalize his movements, Salam focused his objectives on specific regions, making extended visits. 1979 began with such a trip covering California and Mexico; in 1980 came South America, covering Brazil, Peru, Columbia and Venezuela; in 1981 there was a major tour of India, and later the Gulf States – Abu Dhabi, Kuwait, Qatar, Bahrain, Oman, and finally Jordan; in 1984 Africa – Kenya, Tanzania, Uganda, Zambia, Ethiopia, Malawi and Zaire; in 1986 South Asia, covering Pakistan, Bangladesh, India, Malaysia, Singapore, Sri Lanka and Vietnam. 1987 – the year of Salam's bid to head UNESCO – began with an extended trip to (mainly) French-speaking Africa, covering Senegal, Niger, Mali, Ivory Coast, Benin, Nigeria, Togo, Gabon, Cameroon, Zaire, and the Congo. This was followed by a circuit covering Spain, the USSR, Syria, an extended tour of the USA, Canada, Jamaica, Mexico and Argentina, then the UK (including a meeting of Prime Minister Margaret Thatcher's Policy Unit), proceeding on to Tanzania, the Netherlands, the Congo (via the Ivory Coast, Mali and Cameroon), Sweden, the UK, the USSR, China, Pakistan, Austria, France, and finally Pakistan again.

During his 1981 Indian trip, Salam visited Bombay, Madras, Bangalore, Delhi, Calcutta, Bhubaneshwar and Amritsar, as well as a side trip to Bangladesh. A highlight was tea with Prime Minister Indira Gandhi at her unpretentious Delhi residence on 27 January. Salam told her of the friendship and warmth he had met everywhere in India, a contrast to his excommunicated status in Pakistan. Indira Gandhi told him of her continual desire to bring the two nations of the subcontinent together, a process that had been initiated with Pakistani President

Ali Bhutto at a strained meeting in Simla in 1972. Whatever had been prepared at Simla was blown away when both major players fell victim to macabre plots: Bhutto was hanged as a traitor in 1979, while Indira Gandhi was assassinated in 1984.

As a scientific ambassador of the Third World, Salam continually had meetings at the highest level. Zufrullah Khan had first schooled him in the necessary protocol. But at the same time, Salam was not haughty and distant, and remembered his own humble beginnings. On his 1979 trip to Pakistan after receiving the Nobel Prize, he asked his official car to cruise round the campus of Government College, Lahore. Approaching a group of workers, he got out of the car and embraced one of them. It was Saida, a hostel mess servant who had ensured that Salam had been kept supplied with food and drink during his self-imposed solitary confinement while preparing for his college exams. After Pakistan, Salam proceeded to India, where he sought out several of the Hindu and Sikh teachers who had taught him in Jhang.

As a Pakistani and a British Commonwealth citizen, Salam initially had a special immigration line to head for whenever he returned to the UK. However, as Britain increasingly turned away from its imperial past and looked towards a new role as a European nation, the status for Commonwealth immigration at UK points of entry was steadily downgraded. Returning to the UK, Salam now had to aim for the 'aliens' entry line at London's Heathrow airport. Waiting in a rumpled suit alongside forlorn or wretched travellers to face arrogant immigration officials could be unpleasant. Such encounters tested Salam's patience. Frustrated by continual hassles with UK immigration after his globe-trotting, in December 1985 Salam complained to David Mellor, Under-Secretary of State at the UK Home Office (Interior Ministry). Mellor provided him with an official letter to present on return to the UK, and advised Salam to head henceforth for the fast-track entry for European Community citizens.

Salam's family life had to be sandwiched into the available time. However, in the early years, this was not too difficult. His eldest daughter Aziza fondly recalls punting on the river with him in Cambridge. When the family moved to London, she would visit him in his office at Imperial College on Sunday, when they would perhaps go to the nearby Science Museum, a perennial children's favourite. Salam was particularly happy when Aziza became a scientist, a successful biochemist. Later, with Salam so often in transit, and timesharing two marriages, his family life had to suffer. His eldest son Ahmad, living in London, was only two years old when Salam had met Louise Johnson. Ahmad remembered how little quality time he had with his father, often restricted to journeys to and from the airport and to being ferried between the family home in Putney and Imperial College in Kensington. Salam's time was a commodity that could not be squandered and that wealth could not acquire. While television and cinema were deemed a waste of time, reading filled the hours of travel. Whenever he got the chance, wherever he was, Salam would go hunting for books, which returned with him in battered suitcases. His rooms in London and Trieste were stacked with literature – classics, biographies, and dictionaries, but particularly Islam and its culture, and a bizarre collection of self-help books (ballroom dancing, air navigation, ...). He also acquired prayer books and texts from major religions. Classical music was also available, but above all Holy Qur'an recitations were a continual source of inspiration.

Salam gave an initial impression of being smartly dressed, often wearing a dark suit, whatever the weather. Outdoors he invariably wore one of a large collection of hats. After Gieves and Hawkes of London's Savile Row had once made him a suit quickly for his inaugural professorial lecture, Salam patronized them ever after. With so much travel in cramped seats and meals in transit, these suits were not always in a pristine state. Most photographs show his breast pocket bulging with pens and spectacles. Living before the era when physical fitness was deemed to be important, Salam ate a lot of the wrong things when travelling, particularly sweets, biscuits and nuts, which he would keep in his bags and nibble throughout the day, claiming this gave him the mental energy he needed for his work. Close to, his breath wafted the fragrance from spice seeds that he would take from his pocket and chew.

The title of Salam's talk 'Physics and the Excellences of the Life it Brings' at the 1986 physics history meeting 'From Pions to Quarks,' reflected a personal satisfaction. The quote was from Oppenheimer[17]. According to Salam, Oppenheimer had other things in mind than simply the joy of scientific discovery. There was also 'The opportunity physics affords to come to know internationally a class of great human beings whom one respects not only for their intellectual eminence but also for their personal human qualities. In addition [Oppenheimer] had in mind the opportunities that physics uniquely affords for involvement with humankind'[18]. Science was the motive force that drove

Salam's interaction with his fellow men. As well as solving mathematical problems, it opened up avenues for development and for enhancing the human ideal.

In that talk, Salam said that knowing Paul Dirac had been one of the excellences in his life, and that Dirac 'represented the highest reaches of personal integrity of any human being I have ever met'. If Salam modelled his approach to science on Dirac, his personality could not have been more different. Dirac's taciturnity was a legend, Salam was as his most ebullient in science. In 1977, on the fiftieth anniversary of his initial paper on electrodynamics, Dirac had been invited to give a talk at a European conference in Budapest, Hungary. So was Salam. Astute and less hard-pressed conference delegates had managed to get Hungarian visas beforehand and strolled through immigration control at the airport. Others had to be patient while their passports were tediously processed. Among them were Dirac and Salam. During the lengthy wait, Dirac hardly said a word, but Salam's continual laughter resounded across the airport lounge.

To conclude his 'Excellence' presentation, Salam recounted what according to him was the first meeting between Dirac, the consummate introvert, and another giant of modern physics, Richard Feynman, a flagrant extrovert. It had taken place at the Solvay meeting in 1961. (The International Solvay meetings in physics and chemistry, held in Brussels, founded by the Belgian industrialist Ernest Solvay in 1912, were the major focus for physics discussion and development in the first half of the twentieth century.) Salam related 'I was sitting next to Dirac when Feynman came and sat down opposite. Feynman extended his hand to Dirac and said "I am Feynman". Dirac extended his hand and said "I am Dirac". There was silence, which from Feynman was rather remarkable. Then Feynman, like a schoolboy in the presence of a master, said "It must have felt good to have invented that equation". And Dirac said "But that was a long time ago". Silence again.' The story was repeated by Salam in what was intended to be a 'Festschrift' presentation volume to mark Dirac's 80th birthday[19], but which appeared only after Dirac's death in 1984, and again in his talk 'A Life of Physics' given at the 25th anniversary of the Trieste centre[20]. Elsewhere, Dirac and Feynman are described as having first met at a meeting on the future of nuclear science organized in 1946 as part of Princeton University's bicentennial. In 1933 Dirac had suggested a possible link between classical and quantum mechanics that Feynman had seized upon, made it into a mathematical statement and used it in his new treatment of electrodynamics. During a break in the meeting, Feynman asked Dirac if he had known that the two quantities were closely related mathematically. 'Are they?' replied Dirac, who after a silence walked away[21]. Nevertheless, Salam's anecdote of the 1961 Solvay meeting was a good one. First meeting or not, this 'pinteresque dialogue' between Dirac and Feynman was relayed by James Gleick in his biography of Richard Feynman, but where the report of the conversation is attributed instead to Abraham Pais[22].

Speaking at a physics meeting in Puri, India, in 1996, Michael Duff, now Abdus Salam Professor of Theoretical Physics at Imperial College, said 'Theoretical physicists are, by and large, an honest bunch: occasions when scientific facts are actually deliberately falsified are almost unheard of. Nevertheless, we are still human and consequently want to present our results in the best possible light when writing them up for publication. I recall a young student approaching Abdus Salam for advice on this ethical dilemma: 'Professor Salam, these calculations confirm most of the arguments I have been making so far. Unfortunately, there are also these other calculations which do not quite seem to fit the picture. Should I also draw the reader's attention to these at the risk of spoiling the effect or should I wait? After all, they will probably turn out to be irrelevant.' Salam replied: 'When all else fails, you can always tell the truth'.'[23]

In September 1975, a series of lectures at the University of Stockholm on Human, Global and Universal Problems opened with a talk by Salam – 'Ideals and Realities', highlighting the continuing rift between rich and poor nations. The title was inspired by a book *Ideals and realities in Islam*[24], and a written version was subsequently published in September 1976 in the *Bulletin of Atomic Scientists*. As the reputation of his Trieste centre grew, Salam was frequently invited to speak on such topics, and his prose, measured and stately, was very different to the breathless jargon of his scientific papers.

In 1984, a selection of these articles, covering the relevance of modern science to Third World development, and for Islam, together with some scientific talks, and press cuttings eulogizing Salam and his work, were pulled together into a book under the title *Ideals and realities*.[25] It includes some of Salam's compelling speeches advocating the establishment of the Trieste centre and charting its early history. Salam had travelled the world giving these talks, and while the book brought his message to a wider audience, it meant he had to find fresh topics to talk about on his continuing travels. The book had minimal editorial intervention and consequently its contents include a lot of repetition, which is acknowledged in the editors' introduction – 'substantial editing would reduce much of the flavour and emphasis of the original articles'. The book ran to several editions, and was translated into Arabic, Bahasa (Indonesian), Bengali, Chinese, French, Hausa (Nigeria), Hindu, Italian, Japanese, Malay, Persian, Portuguese, Punjabi, Romanian, Russian, Spanish, Swahili, Tamil, Turkish and Urdu. Salam's efforts to promote science in Islam were highlighted in another repetitive collection *Renaissance of science in Islamic countries*, published in 1994[26]. The problems facing the Third World were the subject of 'The South Commission' set up in 1986 by Julius Nyerere, former President of Tanzania. Salam was a founder member and chaired the working party on science and technology. His 1988 report – known at Trieste as 'The Red Book' – underlined the futility of armaments expenditure: while military research and development work can lead to useful spinoff, it is not an efficient way to assure scientific progress.

The contents of these various anthologies overlap, as do the contents of the individual papers, even inside the same publication. But repetition of key messages is a technique widely used in the Holy Qur'an. Salam had a fund of anecdotes and examples that he would often use to illustrate the points he made. The story of Salam's alter ego, Michael the Scot, turned up many times. Another was that of an obscure Islamic physician Al Asuli, from Bokhara, who compiled a dual medical pharmacopeia: 'Diseases of the Rich', and 'Diseases of the Poor', that Salam used to illustrate his message about modern science and technology.

Salam's talks were inevitably coloured by their subject matter. Most of the time, he would have to cover the colourless world of invisible particles interacting through incomprehensible mechanisms. Understanding the mathematical rigour, he would not stoop to inaccurate analogies. He envied poets and writers who could reach out to their readers and instantly enrich their world. In the 1988 Faiz Memorial Lecture in Lahore[27], he confessed to being humbled by the inability to communicate the beauty of his own work[28].

Salam was a passionate man, and this passion had many facets. In the Muslim society in which he grew up, women were shrouded and distant. For a Cambridge undergraduate in the middle of the twentieth century, there were few women visible. And the world of subnuclear physics in which he was immersed was dominated by men. It was only when Salam came to London that he was exposed to emancipated femininity. It was a world for which he did not know the rules, and that were anyway changing fast. Soon, women began to flaunt their feminism. While Muslim women do not display any skin or hair to strange men for fear of precipitating sexual arousal, in London in the early 1960s women brazenly revealed their thighs in the street. This stirred Salam, and he understood why the Holy Qur'an instructed women to show their beauty only to their family and to 'male attendants who have no sexual appetite'[29]. His attractive secretary at Imperial College wore a microskirt with high heels and reported that, when summoned to his office, Salam stared at her legs. In Trieste, he enjoyed the voluptuousness of Italian women of all ages, not just teeny-boppers following fashion. Whenever he entered a room, he noticed women. As his reputation and stature grew, women noticed his aura of brilliance and his animal magnetism. His eyes gleamed with curiosity. He enjoyed it when they noticed. The sudden convulsion of prurience in the 1960s caught many men by surprise. But for Salam's new experience of metropolitan life, it became the norm, and he adapted to it as smoothly as he had absorbed post-war British lifestyle when he arrived at Cambridge in 1946, increasing a cultural versatility that was already highly developed.

In the early 1960s, these influences led Salam to take an ambivalent attitude to the demands of his religion, and he became adept at moulding them to his convenience, rather than submitting himself to their strict interpretation. At Imperial College, where research colleagues had noticed the bottle of Scotch in his office, he often ate lunch at the cafeteria, where one ubiquitous food item was English pork pie. Although technically forbidden to Muslims, this delicacy tempted Salam. The dilemma was overcome by Salam pointing to the pie and then asking co-researcher Gordon Feldman 'what is that food?' Feldman knew that the reply required of him was 'beef pie', as Salam felt that the prefatory lie thereby absolved him of sin. This is underlined by Steven Weinberg who relates '[Salam] once told me there was one thing he held against the Jews'. (At this point, Weinberg, himself of Jewish ancestry, was worried what was coming.) 'The prohibition against eating pork,' continued Salam, 'which you have passed down to the Muslims'[30].

However, in 1962, after his *Umrah* pilgrimage, a zealous Salam returned to piety, and stayed there. Whenever he could, he attended major gatherings of the Ahmadi Community, where passionate speeches about the Prophet Muhammad and on Mirza Ghulam Ahmad, the Promised Messiah of the sect, could bring Salam to tears. Although his faith became unshakeable and survived excommunication, his own religious observance was unobtrusive: he managed to pray every day without apparently interrupting his other responsibilities. In fact his objectives fed on each other – no matter how deeply knowledge penetrates, it will never attain the ultimate level, and there is no reward. The Holy Qur'an, after the opening reverence to Allah, says:

'This is the Book,
Wherein there is no doubt,
A guidance to the God-fearing,
Who believe in the Unseen,
Beyond the reach of human perception.'[31]

In his speech in a UNESCO meeting on 'Islam and Science' in 1984 and dedicated to the memory of his father[32], Salam pointed to this message, 'the timeless spiritual message of Islam, on matters on which Physics is silent, and will remain so'.

There were three columns of ambition supporting the platform of Salam's life:

1. to make fundamental discoveries in science and understand better Allah's work;
2. to enable isolated researchers from developing countries to interact and become intellectually energized without having to emigrate; and
3. to improve the status of modern science in Islamic countries.

How successfully or otherwise did he achieve these three ambitions? For the first, he earned a share of a Nobel Prize. His main scientific contributions – the rationalization of renormalization theory in the early 1950s, the implications of a massless neutrino in 1956, and the unification of weak and electromagnetic interactions in 1968, would and did get solved by others. If Abdus Salam had not been, then parallel work would have ensured that the ultimate outcome of these developments would have been the same, on about the same timescale. Salam's 'The covariant theory of strong interaction symmetries' was launched in 1965 with much trumpeting, but sank virtually without trace; in the 1970s, his effort with Jogesh Pati to forge a grand unification of strong, weak and electromagnetic interactions was off target, but left its mark on what came later. Paradoxically, this errant theory could have been his most influential contribution to science.

For the second objective, it would have been difficult for Salam to have done any better. His International Centre for Theoretical Physics (ICTP) in Trieste become a role-model for many other such international centres. If imitation is the sincerest form of flattery, ICTP's success is easy to see. Its main role is to foster science, not to do it. ICTP is not associated with major research breakthroughs or discoveries, and is not in the same academic league as Caltech, MIT, Imperial College, or Cambridge. But that is not its purpose. It is not a university, it is an adjunct to the international intellectual community, a pacemaker. Nevertheless, Salam's status at Trieste as a scientific figurehead was important in attracting research talent. This was why Budinich was insistent that Salam took up the job as its Director. The Nobel prize in 1979 marked a watershed in the Centre's development, bringing increased visibility, funding and recognition. However, one Salam oversight was in not grooming his own successor at Trieste, an accusation that can also be made of Pakistan's founding father, Muhammad Ali Jinnah.

His third objective, to improve the status of modern science in Islamic countries, blazed continually in his mind. Salam maintained that Islam should strive to align itself and keep pace with modern science. In his introduction to Pervez Hoodbhoy's book *Islam and science*[33], Salam said 'There is no question but today, of all the civilizations on this planet, science is weakest in the lands of Islam'. Throughout his life, almost every major speech recalled the golden age of Islamic science that had kept burning the torch of Greek culture and eventually passed it to Renaissance Europe. Salam used many platforms to urge a rekindling of interest in science in Islamic countries[34]. Tirelessly he urged the oil-rich lands of West Asia and further afield to fund science and technology. Occasionally, there were sparks of interest. Briefly, during the regime of President Ayub Khan, Salam had helped set Pakistan country on a firm scientific path, but in the continual turmoil of that nation's politics, subsequent leaders were less far-sighted or had other objectives, and Salam's message languished and became inaudible. Pakistan may now have its atomic bomb, but the remainder of its science and technology is in a lamentable state, especially when compared to its neighbour, India.

Outside Pakistan, his efforts were channelled into schemes for an *Ummat-ul-Ilm* (Community of Science), and an Islamic Science

Foundation to promote modern science in Islamic countries. In 1990, the ICTP received a quarter of a million dollars earmarked for Arab scientists from the Kuwait-based Arab Fund for Economic and Social Development. But otherwise these words went unheeded. As a marriage broker between Islam and science, Salam was following an unfortunate tradition. The philosopher and scientist Al Kindi (801–873) was the intellectual touchstone of ninth-century caliphs until a periodic resurgence of orthodoxy condemned him to a public flogging. A hundred years later, even the mighty figure of Ibn Sina (Avicenna) had to flee persecution and seek refuge. After another century had passed, Ibn Rashid (Averroës) had to leave Muslim Spain in disgrace.

Antipathy between modern science and religious orthodoxy runs deep. Galileo was excommunicated and Giordano Bruno burnt at the stake for saying that the Earth revolves round the Sun. Darwinian evolution continues to be controversial. Salam's electroweak picture was denounced as heretic Sufism[35]. But in Islam, modern science has had a particularly hard time. When science surged forward in renaissance Europe and continued to race ahead in the industrial revolution, it dimmed on the radar screen of Islam, which retained a frozen image.

Paralleling Edward Said's orientalism is a reciprocal 'occidentalism', in which those in the East view the West through another lens that limits and distorts the perceived image. With the overall landscape hidden, the two standpoint telescopes of East and West peer directly at each other, each vainly trying to discern the virtual image perceived by the other. These shadowy blurs, like dim images from the depths of outer space, are easy to misinterpret. Those few like Abdus Salam and Said, straddling cultures and continents, can peer above the fog of prejudice and try to see a fuller picture. From his fragile Ahmadi foothold, Salam rarely spoke out, but when he did, he scorned what he saw:

'In most Islamic countries, a class of nearly illiterate men have, in practice, habitually appropriated to themselves the status of a priestly class without possessing even a rudimentary knowledge of their great and tolerant religion. The arrogance, rapacity and low level of common sense displayed by this class, as well as its intolerance, has been derided by writers and poets. This class has been responsible for rabble-rousing throughout the history of Islam.... I have been asking the *ulema* (priesthood) why their sermons should not exhort Muslims to take up science and technology, considering that one-eighth of the Holy Book speaks of *taffaqur* and *tashkeer* – science and technology. Most have replied that they would like to do so but do not know enough modern science'.[36]

The failure of Salam's ambition for Islamic science, whether in Pakistan or further afield, can be attributed to a single cause: his isolation from mainstream Islam because of his Ahmadi status. But even this tragedy was reinforced, even overshadowed, by Salam's terminal illness. The story is redolent of that of the Book of Job, where an affluent landowner becomes the victim of a series of horrific afflictions after God makes an enigmatic pact with Satan. Throughout his suffering, which lasted almost all his life, Job never lost faith and remained totally obedient, for only in this way could he fight the nightmarish tortures heaped upon him. The parable shows how even perfect loyalty and fervent belief cannot prevent tragedy. The story illustrates the eternal conflict between carnal weakness and moral integrity and has been invoked by puzzled Jews to 'explain' the Holocaust. To underline human spiritual and physical fragility, the Book of Job poetically compares futile individual efforts to the seemingly unlimited power of science and technology, even as it was known in Biblical times.

Salam, like Job, struggled indefatigably against obstacles but remained obdurately obedient to God, despite repeated failure and humiliation. In a tribute at the 1997 Memorial meeting at Trieste, the Pakistani physicist and communicator Pervez Hoodbhoy compared Salam to the mythological Greek figure of Sisyphus, allegedly the most cunning of mortals. After Sisyphus squealed on Zeus after the King of the Gods had abducted Aegina, the daughter of the River God Asopus, Zeus struck Sisyphus with a thunderbolt, hurling him into the underworld, where he was condemned forever to roll an enormous rock uphill. Each time Sisyphus paused in his efforts, so the rock would roll back and he would have to start again.

Salam was a great scientist, as his Nobel Prize attests. He was greater as an organizer and achiever, but greatest as a voice, a conscience of justice, speaking for the advancement of science among the disinherited. Born deep within a part of humanity that had little birthright, by the sheer force of his own ability he pulled himself onto the high ground of human achievement. As he made this ascent, he looked down and saw the gross inequity in the world where he had come from, and this angered him. The full force of this fury was never vented at one person or any single obstacle. Instead it was a cold, nagging anguish, a resentment of the accumulated discrimination against those parts of mankind whose own potential lies unguessed through a barrier to opportunity. It is an obstacle that perpetuates the injustice against those already underprivileged.

In the *Seven pillars of wisdom*, T.E. Lawrence ('of Arabia') wrote 'All men dream, but not equally. Those who dream by night in the dusty recesses of their minds wake in the day to find that it was vanity: but the dreamers of the day are dangerous men, for they act out their dream with open eyes, and make it possible. This I did.' So did Salam. Once released from the cocoon of his modest beginnings, his relentless drive, his 'cosmic anger', continued for the rest of his life, but frequently had to combat blindness, prejudice and disinterest. Here and there some seeds of this anger fell on fertile ground, and flourished. Salam left other seeds that have yet to germinate. As in the dry desert that blooms after a sudden rain following years of drought, perhaps these desiccated husks can still flower. As years pass, those who visit the institute in Trieste or learn of the science that bears Salam's name will know less of him as a man, but his pioneer aspirations live on as his spirit is fervently passed from one generation to the next.

REFERENCES

1. Budinich, P., *L'arcipelago delle Meraviglie*, (Rome, Di Renzo, 2000)
2. Zewail A., in *100 reasons to become a scientist*, ICTP, Trieste, 2004
3. http://www.middle-east-online.com
4. Dalrymple, W., *The last mughal* (London, Bloomsbury, 2006)
5. Runciman, S., *The first crusade* (Cambridge, CUP, 1980)
6. Said, E., *Orientalism* (London, Routledge and Kegan Paul, 1978)
7. Pakistan Daily Times, Editorial, 22 November 2006
8. see http://www.alhafeez.org/rashid/ludhianvi/abdussalam.html
9. http://www.irshad.org/brochures/absalam.php
10. Sura 2.275
11. http://www.irshad.org/brochures/absalam.php
12. From a total of 45 in 28 countries
13. Crease, R., Mann, C., *The second creation, makers of the revolution in twentieth-century physics* (New York, Macmillan, 1986)
14. From an article in the Urdu monthly magazine 'Tahzeebul Akhlaq', Aligarh Muslim University, India, January 1986. http://www.alislam.org/library/links/00000126.html

15. Gleick, J., *Genius, Richard Feynman and modern physics,* (New York, NY, Random House, 1992)
16. Gleick (1992)
17. Salam, A., *Physics and the excellences of the life it brings*, in *Pions to quarks*, Brown, L., Dresden, M., Hoddeson, L., (ed.) (Cambridge, CUP, 1989)
18. Salam, A., *'Excellences'* (1989)
19. Kursunoglu, B., Wigner, E., (ed.) Reminiscences about a great physicist (Cambridge, CUP 1987)
20. Salam, A., *A life of physics*, in Cerderia, H. A., Lundqvist, S. O., (ed.) *Frontiers of physics, high technology and mathematics* (Singapore, World Scientific, 1990)
21. Gleick, (1992)
22. Pais, A., *Inward bound* (Oxford, OUP, 1986)
23. http://feynman.physics.lsa.umich.edu/~mduff/talks/1996%20-%20A%20Tribute%20to%20Abdus%20Salam/salam.html
24. Nasr, S. H., *Ideals and realities in Islam*, (London, Allen and Unwin, 1966)
25. Hassan, Z., Lai, C. H., (ed.) *Ideals and realities, selected essays of Abdus Salam*, (Singapore, World Scientific, 1984.)
26. Salam, A., (ed. Dalafi, H. R., Hassan, M. H. A.) *Renaissance of science in Islamic countries*, (Singapore, World Scientific, 1994).
27. Faiz Ahmad Faiz (1911–84) was an Urdu bard in the tradition of Ghalib. Born in the Punjab, he remained in Pakistan after partition. A staunch Marxist, he was especially remembered in the then Soviet Union.
28. Zainab Mahmood, http://www.chowk.com
29. Sura 24, v31
30. http://www.forward.com/articles/11003/
31. Sura 2; 2,3
32. Salam, A., (ed. Dalafi, H. R., Hassan, M. H. A.) (1994)
33. Hoodbhoy, P., *Islam and science*, (London, Zed Books, 1991)
34. Salam, A., (ed. Dalafi, H. R., Hassan, M. H. A.) (1994)
35. Vauthier, J. *Abdus Salam, un physicien* (Paris, Beauchesne, 1990) 67
36. Salam, A., foreword in *Islam and science*, Hoodbhoy, P., (London, Zed Books, 1991).

Bibliography

This includes most articles and books referred to in the footnotes, together with other useful sources. Websites, papers in learned journals, and some obscure items have been indicated only in the footnotes.

HISTORY AND SETTING

Lane Fox, R., *Alexander The Great* (London, Allen Lane, 1973)
Gilbert, M., *Second World War* (London, Weidenfeld and Nicholson. 1989)
Morris, J., *Trieste and the meaning of nowhere* (London, Faber and Faber, 2001)
Polkinghorne, J., *Beyond science – the wider human context*, (Cambridge, CUP, 1996)

THE INDIAN SUBCONTINENT

Ayub Khan, M., *Friends not masters* ((Oxford, OUP, 1967)
Bhutto, B., *Daughter of the East* (London, Hamish Hamilton, 1988)
Collins, L., Lapierre, D., *Freedom at midnight* (London, Collins, 1975)
Dalrymple, W., *The age of Kali* (London, Harper, 1998)
Dalrymple, W., *The last Mughal* (London, Bloomsbury, 2006)
Ferguson, N., *Empire* (London, Allen Lane, 2003)
Fishlock, T., *India file*, (Delhi, Rupia, 1983)
Gualtieri, A., *The Ahmadis* (Montreal, McGill-Queen's University Press, 2004)
Hardy, P., *The Muslims of British India* (Cambridge, CUP, 1972)
Man, J., *Genghis Khan* (London, Bantam, 2004)
Marozzi, J., *Tamerlane* (London, Harper Collins, 2004)
Nehru, J., *Toward freedom* (Oxford, OUP, 1936)
Shaw,I., *Pakistan* (London, Odyssey, 1996)
Spear, P., *History of India*, Volume 2 (London, Penguin, 1990)
Thapar, R., *Early India* (London, Allen Lane, 2002)
Ziring, L, *Pakistan and the 20th century* (Karachi, OUP, 1997)

MATHEMATICS, SCIENCE AND BIOGRAPHY

Cropper, W., *Great physicists* (Oxford, OUP, 2001)
Eden, R.J., Polkinghorne, J. C. *Dirac in Cambridge*, in *Aspects of quantum theory*, ed. Salam, A. and Wigner. E., (Cambridge, CUP, 1972)
Gleick, J., *Genius, Richard Feynman and modern physics*, (New York, NY, Little. Brown, 1992)

Hardy, G. H., *A mathematician's apology*, (Cambridge, CUP, 1992)
Hoyle, F., *Home is where the wind blows,* (Mill Valley, CA, University Science Books, 1994)
ICTP Trieste, *100 reasons to be a scientist*, (Trieste, ICTP, 2004)
Johnson, G., *Strange beauty, Murray Gell-Mann and the revolution in 20th -century physics* (London, Jonathan Cape, 1999)
Kanigel, R., *The man who knew infinity, a life of the genius Ramanujan*, (New York, Scribners 1991)
Kronig, R., Weisskopf, V. (ed.), *Collected scientific papers of Wolfgang Pauli*, (New York, NY, Wiley-Interscience, 1964)
McManus, C., *Right hand, left hand*, (London, Weidenfeld & Nicholson, 1992)
Mitton S., *Fred Hoyle, a life in science* (London, Aurum Press, 2005)
Enz, C., *No time to be brief—a scientific biography of Wolfgang Pauli*, (Oxford, OUP, 2002)
Newman, H., Ypsilantis, T., (ed.) *History of original ideas and basic discoveries in particle physics*, (New York, NY, Plenum, 1996)
Watson, A., *Soldier, Scientist and Statesman – a biography of Yuval Neéman* (Tel Aviv, Ramot, 2006)

ISLAM, RELIGION AND THE ORIENT

The Holy Qur'an, Arabic text with English translation (Hockessin, Delaware, Noor Foundation, 1990). Salam cited the translations by A. J. Arberry (Oxford, OUP, 1957) and M. Pickthall (New York, NY, Everyman's Library, 1937).
Hoodbhoy, P., *Islam and science* (London, Zed Books, 1991) Preface by Salam
Maalouf, A., *The Crusades through Arab eyes* (London, Al Saqi Books, 1984)
Mango, A., *Atatürk* (London, J Murray, 1999)
Robinson, F. (ed.), *Cambridge history of the islamic world* (Cambridge, CUP, 1996)
Rogerson, B., *The Prophet Muhammad* (London, Little, Brown, 2003)
Runciman, S., *The First Crusade* (Cambridge, CUP, 1980)
Said, E., *Orientalism* (London, Routledge and Kegan Paul, 1978)
Smart, N., *The world's religions* (Cambridge, CUP, 1989)
Wheatcroft, A., *Infidels – a history of the conflict between Christendom and Islam* (London, Viking, 2003)

PHYSICS

Brown, L., Dresden, M., Hoddeson, L., (ed.) *Pions to quarks*, (Cambridge, CUP, 1989)
Brown, L., Dresden, M., Hoddeson, L., (ed.) *The rise of the standard model*, (Cambridge, CUP, 1997)
Cao, T. Y., *Conceptual developments of 20th century field theories* (Cambridge, CUP, 1997)
Crease, R., Mann, C., *The second creation, makers of the revolution in twentieth-century physics* (New York, Macmillan, 1986)

Dirac, P. A. M., *The principles of quantum mechanics*, (Oxford, OUP, 1958)
Fraser, G., (ed.) *The particle century*, (Bristol, IOP Publishing, 1998)
Ne'eman, Y., Kirsh, Y., *The particle hunters*, (Cambridge, CUP, 1986)
Pais, A., *Inward bound* (Oxford, OUP, 1986)
Riordan, M. *The hunting of the quark* (New York, NY, Simon and Schuster, 1987)
Salam, A., Fonda, L., (ed.), *Contemporary physics: Trieste Symposium 1968* (2 vols), (Vienna, IAEA, 1969)
Salam, A., *The unification of fundamental forces, 1988 Dirac Lecture*, (Cambridge, CUP, 1990)
Weinberg, S., *The quantum theory of fields* (3 Vols) (Cambridge, CUP, 1995)

SALAM

Ali, A., Isham, C., Kibble, T., Riazzudin (ed.) *Selected papers of Abdus Salam* (Singapore, World Scientific, 1994)
Ali, A., Ellis, J., Randjbar-Daemi, S. (ed.) *Salamfestschrift* (Singapore, World Scientific, 1994)
Budinich, P., *L'arcipelago delle meraviglie*, (Rome, Di Renzo, 2002)
de Greiff Acevedo, A., *The International Centre for Theoretical Physics 1960–1979: Idealogy and Practice in a United Nations Institution for Scientific Cooperation and Third World Development*, Dissertation submitted for the Degree of Doctor of Philosophy, University of London, Imperial College of Science, Technology and Medicine, December 2001
Ghani, A., *Abdus Salam: A Nobel Laureate from a Muslim country*, (Karachi, published privately, printed Ma'aref, 1982)
Hamende, A. M. (ed.), *Tribute to Abdus Salam*, (Trieste, ICTP, 1999)
Hamende, A. M., *A guide to the early history of the Abdus Salam International Centre for Theoretical Physics*, (Trieste, Consortium for Development and Research, 2002)
Khidir Hamza, *Saddam's bombmaker* (New York, Simon and Schuster, 2001)
Salam has several walk-on parts
Kibble, T. W. B., *Muhammad Abdus Salam*, Biographical Memoirs of Fellows of the Royal Society 44, 385–401 (London, Royal Society, 1998)
Lai, C. H. (ed.), *Ideals and realities, selected essays of Abdus Salam*, (Singapore, World Scientific). Several editions, with shifting content
Salam, A., *Gauge unification of fundamental forces* Reviews of Modern Physics **52**, 525–38 (1980) The Nobel Prize lecture
Salam, A. *Science sublime*, in *A passion for science*, Wolpert, L., Richards, A., (ed.) (Oxford, OUP, 1988)
Salam, A., *Physics and the excellencies of the life it brings*, in *Pions to quarks*, Brown, L., Dresden, M., Hoddeson, L., (ed.) (Cambridge, CUP, 1989)

Salam, A., *A life of physics*, in Cerderia, H. A., Lundqvist, S. O., (ed.) *Frontiers of physics, high technology and mathematics* (Singapore, World Scientific, 1990)

Salam, A., (ed. Dalafi, H. R., Hassan, M. H. A.) *Renaissance of sciences in Islamic Countries*, (Singapore, World Scientific, 1994)

Salam, S., *Abdus Salam: Science, Islam and Pakistan 1961–74*, History dissertation, Cambridge, 2003

Schaffer, *D., TWAS at 20, a history of the Third World Academy of Sciences*, (Singapore, World Scientific, 2005)

Singh, J., *Abdus Salam, a biography* (New Delhi, Penguin, 1992)

Vauthier, J., *Abdus Salam, un physicien* (Paris, Beauchesne, 1990)

Index

AS refers to Abdus Salam, f to a footnote